More Hot Air

by Tony Kordyban

TABLE OF CONTENTS

Chapter 1.1 The Best Worst Case 9

The requirements say to measure the product temperature under the
"thermal worst case environment." But the Reliability Department, the
Safety Compliance Department, the Thermal Engineer and the Customer
all have different ideas of what the thermal worst case should be.

Chapter 1.2 Blowing the Rel Test 15

The blowers in the Reliability Test Chamber makes the air flow backward
through your chassis. Does that seem like a fair test? Or does it actually
tell you something useful about your product design?

Chapter 1.3 The Five-Finger Thermometer 21

Why your hand is not a good thermal sensor. It not only has calibration
problems, but you might literally get burned.

Chapter 1.4 T-types Fried My Brain 27

There is a reason why different types of thermocouple wire have unique
color codes. You can't always tell the difference between the types by just
using common sense. It takes brains.

**Chapter 1.5 Permutations and Combinations Add Up to
Job Security** 33

Maybe it makes sense to stack shelves up in a rack and use one big fan
box to cool all of them. But there are lots of thermal reasons not to like
that design. Endless combinations of hardware can keep you doing
thermal testing for years.

An important limit to forced air cooling is the audible noise of the fan. As fan RPM goes up, so does the flow, but the noise goes up even faster, according to the Fan Laws.

A talk-radio-show host discovers that while paying attention to component temperature, one can forget that the cooling fan itself is a component, too, and has its own operating temperature limit.

What does it mean for an electronic component to have an operating temperature limit? Does it blow up 1 degree over the limit? Does it slowly degrade in function, or does it use up some of its allotted life? Wouldn't it be nice if the component manufacturer would tell us?

Fuses are easy to ignore, but some very common types need to be de-rated for temperature. Just because they don't dissipate heat doesn't mean they don't get hot.

Can the printed circuit board act as a heat sink for a component? Perhaps, but a story about a naughty boy in a swimming pool explains the practical limits of this idea.

In thermal analysis of a circuit board, you often ignore all the capacitors, because they aren't supposed to add heat, and there are so darned many of them. But capacitors can generate heat, and their properties can shift with temperature.

A baffle is often used to deflect hot air from the exhaust vent of one chassis to prevent it from getting sucked into the inlet vent of another. But a baffle is not a perfect way to isolate neighboring chassis, because heat can conduct through the baffle plate. Maybe changing the plate from metal to plastic will help. Don't count on it.

Chapter 7.5 Normal Room Temperature: the Latest Worst-Case Thermal Condition
Herbie slows fans down to meet the NEBS audible noise limit. But does that make normal room ambient the worst case thermal situation? Not because of the ambient, but because the air flow is the lowest?

Chapter 7.6 The Weakest Link in Air Cooling
It is the 21st century already. We don't have our personal jet packs yet, and we're still struggling to get heat out of a room with air-conditioning. It turns out that air-conditioning technology is fine—it's the puny humans in the equation that are limiting the air cooling. Where are our robot servants to save the day?

Introduction: Everyone Needs a Human Brain Unit

Everything you write is a confession.

You can't help revealing something about yourself in even the most innocuous scribble or note. Your grocery list exposes your weakness for fatty and salty snacks. That e-mail thank-you note to your mother-in-law shows the true depth of your appreciation for the exercise video she gave you for your birthday. And even that seemingly objective engineering test report you wrote says more about how you expected the product to perform than what it actually did.

I will make my confession right up front, so you won't have to infer it from the upcoming chapters. What I really wanted to write was a science fiction novel called *The Human Brain Unit.* It was going to be this really cool, Stephen King–type story. The phone company discovers a small segment of the population that is telepathic. Instead of sharing their discovery with the world, the company turned it into a telecom development project (since that is all they know how to do, anyway.) Their scientists find a way of harnessing the telepathic powers of these relatively rare psychics to replace segments of the telephone network. Instead of sending signals over traditional cables, microwaves, satellite links or optical fibers, they can be instantly communicated from one brain to another. The main hitch in the idea of such a product is that if people could communicate directly with one another telepathically, how could the phone company charge them for it?

Their solution is that they don't replace the entire phone network. Your grandma in Toledo still has to pick up her handset and dial a number and talk into a microphone and listen to a tiny speaker. But after her phone call gets to the central office, it goes through an electro-psychic converter, and gets sent from the human brain unit (HBU) in Toledo telepathically to a corresponding HBU in Fresno, where it is converted back to ordinary electronic signals that reach the phone of your auntie with the gall bladder trouble.

A boring idea by itself, I admit. Most people don't know how the real telephone network works anyhow, so they wouldn't much care if it were replaced by a bunch of brains hooked up to electrodes. Maybe I could have spiced up the story by having the evil phone company snatching peoples' brains and sticking them in glass casserole dishes filled with bubbling green fluid. And later, somehow the network of brains would start to take on a life of their own, grab control of the network and mete out a whole slew of poetic justice to the evil telephone company engineers.

You can see why I never got anywhere with that novel. There just wasn't any way to work in a hot love story when most of the characters are evil engineers (no believable love story there) or disembodied brains.

That leads to another point of confession. I said earlier that most people haven't got a clue how the phone network works. I worked in the telecom industry for nearly 17 years, and I have to count myself among that number. I worked on literally dozens of projects that developed new electronic hardware for the telephone network. I was aware that they generally had something to do with cell phone switching, or conglomerating the phone calls from many lines onto one. But other than that kind of vague notion, I didn't understand what the circuits were supposed to do. As a thermal engineer, I knew they had one thing in common: They converted electrical power into heat, and it was my job to figure out how to get that heat out, so the circuit would not overheat.

When I started to write about my adventures in the newsletter articles that would eventually become this book, there were two problems. First, I had to disguise the project I was writing about, because usually I was writing about some embarrassing thermal design mistake. I didn't want to use real names of people and projects, even if I could get permission from management, because I didn't want to hurt anybody's feelings or reputations. I did want to write about the blunders to share their educational value with others in the hopes that others would avoid those same boo-boos. The second problem was that I didn't understand the real projects well enough to describe them without making myself look stupid.

My answer to that double dilemma was to just make up fictional projects. Instead of the double-density echo-canceling circuit cards and fiber-optic switch matrices I actually worked on, you will find yourself reading about Lost Dog Finding Systems and Telemarketer Disabling Circuits. And time and again, you will find the human brain unit as the backdrop of the thermal lessons.

That's your introduction. The purpose of the introduction is to tell you about things that just pop up out of nowhere in the book with no explanation. The human brain unit is one of those things. I keep referring to it as one of Herbie's projects. Now that I've told you where it came from, it won't be so confusing when you run across it later.

Oh yes, Herbie. If you haven't read my first book, *Hot Air Rises and Heat Sinks,* you don't know who Herbie is. Herbie is my friend. He is fictional. He is an engineering archetype. He is that guy who is not quite as good at thermal design as you are, so you can blame all the thermal mistakes on him. You know somebody like Herbie where you work. He is enthusiastic, gets things done and is willing to work beyond his many limitations.

Herbie is important. If I had not invented him, God would have had to create him. Herbie is a bit thick, but he serves as our teacher, because we learn from his mistakes. Without him, we would have to make these mistakes ourselves.

He also serves as a living warning. "Don't be me," says he, "Learn. Read this book."

Organization of the Book

You should have noticed by now that this is not an engineering textbook. It is not going to start by introducing Conduction, Convection and Radiation, and then have you work out homework problems. This is a collection of short case histories, mostly based on things that really happened, although if I've done my homework, you'll never be able to trace them back to the real people and projects they are based on. There is no logical progression through the book. Each chapter was originally an article in a monthly thermal design newsletter called *HOTNEWS*. I wrote about whatever had caught my attention recently. So if it strikes you that the chapters are somewhat disconnected, congratulations, you are right.

The chapters are loosely organized into seven sections. They were organized in the same way that laundry gets organized after it comes out of the dryer. I pulled chapters out of the pile and held them up next to each other to see what went with what. The chapters that didn't seem to go with any other chapters were thrown together in their own section, like a drawer full of unmatched socks. Maybe those chapters are still useful as sock puppets or something. This is how the piles of chapters are sorted out:

Section 1. Measurement and Test: Getting the Wrong Answer Direct from the Lab

Section 2. Fans: Increasing the Air Flow and the Trickiness of Your Cooling System

Section 3. Components and Materials: the Sum of the Parts Is Sometimes Just a Big Hole

Section 4. Radiation. No, Stefan and Boltzmann were not a '70s German heavy metal band!

Section 5. Tales of the JEDEC Knight – a Crusade Against the Industry Standard Definition for Component Thermal Resistance

Section 6. A Collection of Not Even Loosely Related Stories

Section 7. Telecom: A Field with Myths and Mistakes All Its Own

This last section has a lot of jargon and thermal concepts specifically related to the telecom industry. I lumped them together to allow the readers from the automotive, aerospace and consumer electronics industries to skip them easily. As with the other chapters, though, they are written for the general reader, and even if you can't glean any relevant technical lessons from them, you might find one or two puns to reward your effort.

Now please read, enjoy, learn, share what you find with your colleagues. Just don't base your product thermal design solely on what you read in this book. Check it out with your own analysis and testing! Just because I assert that "everything you know is wrong," doesn't mean that the corollary is true, that "everything I know is right." Remember, if I have learned enough to be able to write two books on this subject, I must have made my share of mistakes. You can count on one or two new ones to be in this book.

That's right, I'm not perfect. I confess.

Section 1

Measurement and Test: Getting the Wrong Answer Direct from the Lab

Test data is supposed to be "truer" than other kinds of information. Sure, it's better to test products than to not test them. But just because a value was measured with the most accurate meter in the world doesn't mean that it is right. The ways of messing up tests and experiments are myriad. The chapters in this section give just a few interesting examples.

Maybe you doubt me. You think that with the properly calibrated machine you can always discover the truth.

They say the camera never lies.

Look at the photo on your driver's license. Is that what you really look like?

THE BEST WORST CASE

I came back from vacation to a flurry of e-mail about the thermal testing of the optical optioner power supply (OOPS) circuit card. A big system-level design review was coming up, and it wasn't clear that the OOPS had passed. Six different people measured temperature six different ways, and although they achieved six different results, they all focused on the input power isolation transformer as the critical component that would make or break the OOPS.

Don from Test Engineering tested a prototype OOPS by using it to provide power to the actual Optical Interface System for which it was designed. He mounted the whole assembly in the factory burn-in chamber, turned the chamber air temperature up to 50°C and measured the air temperature near the OOPS input transformer. The transformer spec says it operates from −45 to +70°C ambient. Air temperature measured near the transformer was 62°C when the ambient temperature was at its *worst case* of 50°C. Don concluded that the input transformer was OK, because its local air temperature was less than 70°C.

Doc Smith did temperature testing to qualify the OOPS for approval by safety agencies. He didn't care how much current the

Optical Interface System actually drew from the OOPS. He needed to see how hot things got when the OOPS was delivering as much current as it possibly could before collapsing (a power supply is often designed to "collapse" when overloaded or short-circuited. Its output voltage drops to nearly zero when a maximum current is reached.) He used a variable load box, instead of the Optical Interface System, to run it right up to the current limit. Then he let it sit, until something either burned out or the temperature stabilized. Doc was checking whether this *worst case* fault condition would cause a safety hazard. He measured the coil temperature of the input transformer to be 125°C.

Will, a technician in the OOPS development department, put a fully-loaded Optical Interface System, complete with the OOPS, into an environmental chamber at 50°C—its *worst case* ambient. After it had soaked at that temperature for several hours, he turned *off* the chamber fans (because the Optical Interface System is cooled only by natural convection, the chamber circulating fans produce some artificial cooling of the components), and let everything come to equilibrium. The coil temperature of the input transformer on one of the OOPSs was 82°C.

Penny, a senior tech in OOPS development, thought Will's worst case wasn't bad enough. The Optical Interface System has two power supplies, for redundancy. If one fails, the other is meant to carry the total load until repairs can be made. She repeated Will's test in the environmental chamber, this time disabling the second OOPS, so that the one being tested had to carry the entire load by itself. The coil of the input transformer increased to 93°C.

John, in Reliability Engineering measured the coil of the input transformer at 118°C. He ran the OOPS in an environmental chamber at 50°C with the fans off, but he drew current from the OOPS with a variable load box. Unlike Doc Smith, he set the load to the maximum rating from the OOPS specification, instead of running it at its current limit. John also ran it at the lowest rated input voltage of 42V, instead of its nominal input voltage of 48V, so that current in the input transformer would be at its *worst case.*

Maureen, an intern in Reliability Engineering, combined John's results with an extrapolation method from a textbook she found in my office. She estimated the effects of 13,000 feet altitude, and of running the OOPS at its current limit. She extrapolated a *worst case* coil temperature of 160°C.

Herbie was trying to coordinate the design review meeting. "Hey Thermal Guy," he e-mailed, "Which one of these is the real worst case? The input transformer has a 130°C coil temperature limit. Does this thing work or not? Some tests say yes, some say no. There will be managers in this design review. I can't ask them to make a decision when I have conflicting test results!"

As you can see, there are worst cases, and then there are worse worst cases.

You'd like to test your product under every possible combination of operating condition. But time and money are both money, so you take a short cut. You come up with some combination of test parameters that are the Worst Case Conditions. If it passes Worst Case, we don't need to run all the other combinations.

Great time saver, isn't it? But how do you decide what the proper worst case conditions are for doing a thermal test? That depends on the purpose of your test.

Don was checking whether the input transformer met its "ambient" temperature limit; his test was a waste of time. Ambient temperature rating is useless in telling if a component is being applied properly. There is no definition of "local ambient" to tell you where to measure, so you don't know whether the ambient near the input transformer is over 70°C or not.

Doc Smith's safety agency test went beyond the OOPS operating range. Safety agencies are not interested in the function or long-term reliability of the input transformer. They want to make sure that if a fault someplace draws the maximum possible current, a fire or shock hazard won't occur. The safety-related worst case doesn't happen unless there is a short circuit, so it shouldn't be used to decide whether components are cool enough to be reliable.

Will and Penny chose a fairly realistic set of bad operating conditions. The OOPS has to supply current for a fully loaded Optical

Interface System, which is the load it is designed for. Will picked two simultaneous bad conditions: a fully loaded system, and maximum air temperature. Penny added a third simultaneous problem: a fault in the redundant OOPS. That doubled the load on the remaining OOPS. Was that fair? Probably. The Optical Interface System needed to keep working, even if it is 50°C and one power supply has failed. That gives the customer time to replace the failed OOPS.

But what happens a year down the road when somebody comes up with a double-density plug-in board for the Optical Interface System, and the whole system draws more power than the current model? The fully loaded shelf today draws only 37 amps, but the OOPS is rated to 50 amps. That is why John tested it at its full, rated load. He is dead sure that if a power supply is rated to 50 amps, someday, somewhere, somehow, somebody will load it to 50 amps and expect it to work. Then he added another simultaneous condition—he reduced the input voltage to the rated minimum: 42V. His combination of conditions is a pretty bad worst case: maximum load, maximum ambient, minimum voltage. But it is a foreseeable condition that is within the operating requirements for the shelf.

Maureen really piled on the conditions. She went to the worst altitude (recall that at high elevations, air is less dense, so air cooling doesn't work as well) in the product specification, and then increased the load to the current limit. Maybe the altitude should be included, but expecting the OOPS to work thermally when it is loaded beyond its rated capacity is going too far. Maureen assumes a misuse of the OOPS, on top of Pike's Peak, in the middle of a 50°C heat wave. This is Beyond Worst Case.

My opinion (and opinion is important here because there is no objective definition for worst case) is that John's test is a pretty good, realistic, worst case for the OOPS. It includes most of the important conditions that contribute to the temperature of the input transformer. We have to remember that if we design the OOPS to meet every conceivable combination of maximum conditions, it would have to be severely overdesigned, and would cost more than it ought to. John's test finds the input transformer coil to be 118°C, and the coil operating limit is 130°C. It is OK.

Picking a realistic worst case is a gamble. You have to figure out the odds of a particular combination of events happening at the same time. From the OOPS example, Penny assumed that three things would happen together: one OOPS would fail in a fully loaded shelf while the ambient was at 50°C. Most of the time both OOPSs are working and sharing the load. Most of the time the ambient is 25°C or less. If an OOPS fails, an alarm tells the customer to replace it. The product specification gives 50°C as the maximum ambient, but not for continuous operation; 50°C operation is supposed to be infrequent, on the order of three days per year. The odds of overlapping these three events seem pretty low.

But the odds are low only if all the conditions that add up to a really bad worst case are independent events.

This is illustrated by a phone call I got from our Optical Doo-Dad Division out on the eastern fringe of the U.S. Judy had an unusual worst case problem. Her new product was slated to go into U.S. telecom central offices. The environmental spec for those offices is published in an industry standard called *NEBS* (*Network Equipment Building Standards,* published by Telcordia). (See Section 7 for more telecom standards information.)

Judy was new to the telecom biz. She wanted me to help her use the temperature limits from *NEBS* to write a temperature requirement for a new Optical Doo-Dad Component (ODDC). I explained that, according to *NEBS,* a central office normally runs about 25°C, because it is an indoor, air-conditioned environment, but under emergency conditions, can go as high as 50°C, such as when the air-conditioning for the building fails.

"But how many central offices at a time can be at 50°C?" she asked. This was important to her, because ODDCs in a whole bunch of central offices would be linked together. The ODDC is sensitive to temperature. If one office in the chain goes to 50°C, the signal running through them would degrade partially, but still get through. But if more than one office in the chain goes to 50°C at the same time, then the signal would degrade so much that the network would lock up like the Tin Man on a scuba dive.

No industry standard controls the environment for groups of central offices. They are assumed to operate independently, more or less. To figure out a real worst case, you have to look at the odds. What are the odds that more than one central office in a network will be at 50°C at the same time? For a simple network of two offices, assuming the probability of a random air-conditioning failure at 1%, the odds of both offices having a simultaneous 50°C ambient are about 1 in 10,000. That sounds low, but is not actually good enough to meet telecom industry standards for network up-time.

Not only that, but air-conditioning failures are not necessarily independent events. Suppose your network is in northeast U.S. What are the odds that there will be a regional heat wave that causes a widespread power failure? It happens once every summer. That single event can cause *all* the offices in the area to go to 50°C. Telecom offices all have battery plants and generators to keep the critical circuits powered up, but they don't always have enough power to keep luxury items like air-conditioning on-line. The phone network is still supposed to work under those conditions, so it has to be considered a realistic worst case. So Judy has to go back to the thermal drawing board.

This is best summed up by Verse 183 from that legendary folk tune "The Engineering Blues." If you don't know the tune, think of the slow bits of Don McLean's "American Pie":

> *Design review is drawing near*
> *Write a test plan nice and clear.*
> *Choose the proper time and place*
> *Understand your real worst case.*
> *For your final thermal test*
> *Best is worst and worst is best.*

BLOWING THE REL TEST

When I was a kid at St. Nicholas School in Buffalo, New York, I believed that teachers gave tests because they hated us and wanted us to suffer. The nuns told us suffering was good for the soul, and their faces hinted that their souls had been done lots of good.

Sister Mary said the purpose of tests was to see what we had learned. My dad (a professor of engineering) explained that tests were necessary to separate the "good students" from the "dummos." Principal Starch once said that test results were a way of telling how well the teachers were cramming our heads with useful things, for example, the crucial distinction between "further" and "farther."

It seems that the purpose of a test depends on your point of view. What makes a test fair depends on what you think its purpose is. For example, if test grades were going to be used to judge a teacher's performance, how come she got to make up the questions and answers?

At TeleLeap a minor brouhaha had erupted between Test Engineering and Hardware Development, and I was called in to add fuel to the fire. I grabbed my Dr. Grashof Junior Thermal Detective Kit and raced to our manufacturing plant just outside of Cyclone City, the scene of the mystery.

The dispute centered not around a dead body, but a dead HBU[1] circuit board. Without apparent reason, it had started to fail in large numbers during the factory Reliability Test. For the Rel Test, a rack full of HBU systems is put in a large, garage-like chamber and run for at least 24 hours at 50°C.

The hardware engineer who had developed that board announced that it was, without any doubt, a *thermal* problem. A diode in the 2V supply was getting hot enough to go into thermal runaway, causing the input fuse to blow. He could reproduce the failure very easily by over-heating the diode with a heat gun (an oversized blow-dryer, whose normal use is to help strip paint off walls). The weird thing was that he had previously tested the HBU board in his lab at 50°C, quite exten-sively, and it had never gone into thermal runaway. So he pointed the finger at Test Engineering: "Your Rel Test is screwed up. It is obviously overheating the module and making it fail. Fix your chamber."

Test Engineering claimed, "Hey! Our chamber has expensive con-trols to make sure it never goes above 50°C. Just look at these digital temperature displays. They can't be wrong if they're digital! The chamber can't be overheating your board. Something must be wrong with your board, because it keeps blowing. Fix it."

Under intense questioning, they revealed to me a crucial clue: the failures had started right after the grand opening of the new-and-improved Rel Test chambers.

"Improved?" I asked curiously.

"Oh, yeah," my Test Engineering guide said, "See these 20-foot-tall closets with garage doors on the front? These are the old Rel Test chambers. Inside are heaters and fans to raise the air temperature up to 50°C. The problem with them was stratification. Because hot air rises, air near the floor was 40°, belt-high it was 50°, and at the top of the equipment rack maybe 60°C. That wasn't a fair test, because the boards at the bottom of the rack weren't getting as hot as the ones at the top. The fans in these old chambers are rated about 300 cubic feet per minute (CFM). To get rid of the stratification in our new chambers, we put in 3,600 CFM blowers."

"And the HBU board fails only in the new chambers?" I asked.

"Yup. It passed the Rel Test in the old chambers for many months."

I was puzzled. These particular HBU boards are cooled by natural convection. It seemed to me that blowers should make components run cooler, not hotter. I had seen that happen before in environmental chambers (see Chapter 3 in *Hot Air Rises and Heat Sinks*).

I did have a theory why the bigger blower might cause a thermal problem. Maybe when the chamber heater coils kicked on, very hot air (much higher than 50°C) blew directly into the inlet baffle of the HBU chassis. To test my theory, we put an HBU system inside one of the new chambers. I put a thermocouple on the troublesome diode, and one in the baffle to measure the inlet air. Then we cranked the chamber controls up to 50°C.

By the next morning the circuit board had failed, but the temperature log showed that the inlet air temperature had never gone above 52°C. So much for my hot coils theory.

We knew the HBU board failed inside the chamber with the blowers on at 50°C. I was curious how it worked under natural convection, the condition it was supposed to work under in real life (or as real as life gets in the hands of our customer). So we opened up the big garage door and ran the HBU system at normal room temperature with the chamber blower and heater off. The board worked fine, and the diode reached a steady-state temperature without any hint of going into thermal runaway. In natural convection, it ended up with a case temperature about 18°C above ambient. That did not seem very dangerous.

The Test Engineer and I brainstormed inside the chamber. We even stood in there with the blowers on, to get a feel for the air flow patterns. There was a lot of air movement, making the paper ID tags tied to the boards flutter and fly around in all directions.

"Is it the hotter air, or is it the high speed causing the problem?" I asked, "Maybe with all the swirling going on in here, air is actually flowing backward through the chassis in some places."

The Test Engineer said, "We could find out. Let's turn on the blowers without any heat."

That idea cracked the case. We shut the garage door and turned on the 3,600 CFM blower. Then we kept our eyes on the diode temperature. To our astonishment, even though no heat was being added to the chamber, the diode started to get hotter. By the time it stopped going up, it was 26°C above ambient, about 8° more than it was with the blower off. The blower somehow interfered with the natural air flow through the shelf. When we shut off the blower, the diode went back down to 18°C above ambient.

Before driving off into the sunset, I reported these conclusions:

1. **The New Rel Test Chambers have a problem.** Their Aeolian[2] air flow has succeeded in making the air temperature uniform within the chamber. The inlet air temperature at any level in the equipment rack never exceeds 52°C. They do not overstress any circuit boards with high temperature air. But somehow the swirling air interferes with the natural air flow patterns through at least some of the chassis, causing some components to get hotter than they would under normal operating conditions. I don't know exactly how it happens, but it could be related to the phenomenon in the movie *Twister*, where the tornado is strong enough to explode barns, rip up picket fences, and toss around cattle and tanker trucks, but it never even messes up the hair of Bill Paxton or Helen Hunt.

2. **So what if the chamber is screwing things up, it found a weakness in the HBU board design!** The high speed air makes the diode go up an extra 8°C, which is enough, at the maximum ambient, to start the diode sliding into thermal runaway. The board meets all of its operating requirements, but 8°C is not a large margin of safety. A robust product should keep a large cushion between itself and anything as destructive as thermal runaway. Fortunately, this board is already being redesigned, and this diode won't be in the new revision.

A Fair Weather Test

So who was right? Was the board design flawed, or was the Rel Test unfair? It depends on the purpose of the so-called Reliability Test.

The question came up again when the Test Engineer asked me how to fix the new chambers to make them more fair. My answer was, "Fair? What do you want the chambers to do?"

- It is called a Reliability Test, or sometimes "Burn-In". The goal of a burn-in process is to stress a system, to accelerate infant mortality failures, so that they happen in the factory instead of two days after the customer receives them. But our burn-in is at 50°C, and lasts, at most, only a few days, which is within the normal operating range of the system, so there is no stress, no failure acceleration, no reduction in infant mortality. So the purpose couldn't be to increase reliability.
- Some folks say the purpose it to prove that each system we build actually works with all its pieces in place at 50°C. If that is the real reason, then we should build the chambers to mimic natural convection, so we don't understress or overstress any component with blowers.
- Development engineers have a reasonable expectation that the Rel Test should not fail any board that meets all its operating requirements. If it is designed to work at 50°C, it shouldn't have to pass a Rel Test worse than that.
- Test Engineering rightly points out that the Rel Test routinely finds many boards that don't work in the system at 50°C, even though they pass all functional tests at room temperature. If it is so valuable to prevent boards like that from getting out in the field, why should we limit the Rel Test to 50°C? Maybe it would be even more effective at 55° or 60°C. Maybe having blowers that stifle natural convection would be a good way of finding weak board designs.
- The most compelling, but least useful, purpose for the Rel Test is that we told our biggest customers that we do it as part of our quality assurance program. We promised them years ago that we would burn-in every system, even though we had no data that it improves quality or reliability or that it weeds out infant mortality. Now we are stuck with it. We can't just stop doing it without offering to do something else, something which would likely cost more. I suppose any chamber that has a meter that reads 50°C on it would serve this purpose.

So one mystery was solved (the overheating diode), but a bigger mystery remains: What is the real purpose of the Rel Test? When I solve that one, I'll start thinking about how to make the chambers more fair.

Notes

1. HBU is a TeleLeap acronym that stands for Human Brain Unit. (It, like TeleLeap itself, is largely fictitious.)
2. Aeolian is a reference to the mythical Greek god of the winds. I wanted to say "Herculean" but since it had to do with air movement, my literary wife insisted it should be Aeolian.

THE FIVE-FINGER THERMOMETER

My profession has done half its job—convincing folks that high temperature is bad for electronics, but not how to measure it. This time it was a phone call from Australia. Previously it was an e-mail from Brussels, and before that, a fax from Lubbock, Texas. I get the same question in four different languages from all over the Tele-Leap Global Marketplace.

"This is your mate Herbie's ole Uncle Reggie, out of the Sales Office in Sydney. I'm working with an installation supervisor, who says he's got a heat problem with a rack of text message amplifiers. I was hoping you could give us some advice on it."

I asked him for a few details—such as what exactly is a text message amplifier (TMA), what the installation looked like, how the weather was (to see if I needed to make a personal site visit). Reggie explained that the TMA converted shorthand text from cell phones into real words ("CU4 dinner B4 6, H8 Cfud" becomes "See you for dinner before 6 p.m. Remember, I don't care for sea food.") Then, while he expounded on the area's climate in colorful and incomprehensible metaphors, I searched my files.

"I have the report in front of me, if you want any specific data," I said. "Maybe you should tell me just what this heat problem is."

Reggie solemnly said, "The installation supervisor said that the customer's site manager mentioned that he thought the TMA units were running a little hot. So the installation chap went over to the rack and checked it himself, and officially agreed that it was, quote, 'running hot', unquote."

"Hmm," I hummed supportively. "That doesn't sound good. But what's happening? Are circuit boards dying? Is the system losing traffic or taking bit errors? What is the room temperature? Is there an air-conditioning failure?"

"It's not as crystal clear as all that," Reggie said. "Early on there were a lot of board dropouts that we blamed on high temperature. But it turned out we had the wrong version of the firmware for south of the equator, 0's and 1's reversed, as it were. Once that was squared away, the thing's started running smooth as silk. Not a single board has dropped out since. The only snag left is this heat problem."

"But if it's running smooth as silk, what heat problem is there? Did anybody measure component temperatures?" I asked.

"Measure?" Reggie said. "The site manager put his hand on the front door of the top TMA shelf, and it felt hot to him. Then our installation fellow did the same thing. The shelf feels hot to the touch. They believe that heat is as bad for electronics as American TV is bad for your brain cells. They're worried that the heat will shorten the life of the units."

"He felt it with his hand?" I said. "Did he use his right hand or his left hand?"

"What? Right or left hand? I don't know."

"It's very important," I said. "Maybe it's different Down Under, but in the States we use our left hands for measuring Celsius and our right hands for measuring Fahrenheit."

There was a long silence, during which I suppose Reggie looked at both of his hands. "You're pulling my leg for a reason, eh?" he said, finally.

"Yeah," I said. "First of all, the hand, or any part of your body, is not good for measuring temperature. And second, the shelf is supposed to feel hot to the touch, even when it is operating normally. Unless—tell me this—when he put his hand against the sheet metal of the shelf door, did it come away with the TeleLeap logo burned into his flesh? You know, like that Nazi guy in *Indiana Jones* when he grabbed the Egyptian bronze gizmo out of the fire. And if he did, could he read any satanic messages hidden in the backward writing?"

"No, no burns or slogans from Beelzebub. He said that when he put his hand on the door, it felt uncomfortably warm."

"Based on the fact that the metal door did not burn his skin, and assuming the guy doesn't have the skin of a crocodile, I'd estimate the door temperature about 50°C or less. That would feel pretty warm to the touch, but that doesn't mean the electronics see any particular stress. Components could easily have surface temperatures as high as 90°C and still be safely within their operating ranges."

Reggie thought about that for a second: "So I can tell them our thermal expert says it's OK for the product to feel hot to the touch? And it won't affect reliability?"

"Tell them this. Feeling the shelf door with your hand isn't accurate enough to tell if you have a problem. It could feel hot and be OK, or feel hot and be in trouble. But based on the results of our temperature testing, *if* they have installed the shelves per the manual, with the proper baffles and so on, and *if* they keep the ambient within the operating range we specify in the manual, then the components will be at a low enough temperature to have good reliability. If they are worried that there is still something wrong, I can suggest some spots for them to take actual temperature measurements, using calibrated temperature sensors, and I can compare it to my lab data…"

"Hold your horses, Mr. Guru!" Reggie said. "Your say-so was all the assurance they were looking for. I don't think we'll be needing to measure anything with sensors. I'll get back to you if we change our minds."

That usually puts an end to the "heat problem." I never heard from Uncle Reggie again. I'm not sure why, but the folks who detect heat problems with the five-finger thermometer never want to measure an actual temperature for me. Maybe they were scared by a doctor's thermometer in their early childhood, and never want to go near a temperature probe again.

There are three reasons why your hand makes a lousy thermometer:

1. **Slow response time.** How many times have you grabbed something off the stove, and by the time your brain registered **too hot!,** you were already burned? It just isn't safe to touch objects to find out how hot they are.

2. **Too difficult to calibrate.** Your skin is exquisitely sensitive to minute temperature changes, even less than 1°C. That's how you can sense how long it was since someone sat on that wooden chair in the library before you. The trouble is you haven't got a clue what the value of that temperature is. Even a trained hand like mine can't guess a temperature better than ±10°C. And the farther away from room temperature you get, the worse the accuracy.

3. **It's the wrong tool for the job.** The nerves in your skin don't sense temperature at all. They actually sense *heat flux.* The nerves are like a traffic reporter that watches which way the heat is flowing and how much. When you pick up a steaming cup of coffee, which is hotter than body temperature, heat flows into your hand. Your nerves see heat flowing inward and say, *That thing is hot.* When you grab an ice cube, heat flows out of your body into the ice and the nerves say, *Hey, cool ice, man!* This usually works for judging relative temperature. But it can easily get screwed up. For example, think back to a typical Illinois spring day, when you've been out shoveling snow for three hours. Your hands are nearly numb from the cold as you wash up for dinner. You stick your hands under the cold water tap. The cold water feels so hot that it seems to burn your skin! The outer flesh of your hands is actually colder than the tap water, so heat is entering your skin and your nerves interpret this as *Danger—hot water!—scream a naughty word or two!*

The same thing happens in this next little experiment. Grab a metal doorknob. It feels cold. Grab the door with the same hand by the wood. The wood doesn't feel quite as cold as the knob. But they are both at room temperature! The metal knob is a good heat conductor, so it easily transports heat out of your skin. The nerves interpret that as cold, or low, temperature. When you grab the wood, a good thermal insulator, very little of your body heat conducts into it. Low heat flow, so the nerves say it's not so cold. What good is a thermometer that gives a different answer depending on the type of material it is measuring?

Which brings me back to the subject of burning the company logo into the palm of a hand. It is not just an illusion that metal feels cooler than plastic, or that your rear end gets colder when you sit on a marble bench than a leather sofa. The thermal conductivity of the material you touch can make a big difference in the temperature it has to be to hurt you. Table 1-1 is derived (stolen) from an electronics industry safety standard. I don't know how they got their data, unless it was from Nazi prison camp experiments, or testing on helpless graduate students. Please follow the real safety standard that applies to your particular product.

This makes sense when you look at the thermal conductivity of the materials (Table 1-2).

Table 1-1

Type of Touchable Part	Maximum Allowable Temperature (°C)		
	Metal	**Glass**	**Plastic**
Handles, knobs touched for short periods only	60	70	85
Handles, knobs continuously touched	55	65	75
External surface, which may be touched	70	80	95
Parts inside, which may be touched	70	80	95

Table 1-2

Thermal Conductivity Ranges

Metals	Glass, stone	Plastic, wood
High	Medium	Low
50–400 W/m/°C	1–5	0.1–0.8

These tables have three uses for the electronics designer:

1. In a rough way, you can measure a surface's temperature in a far off land by calling someone there and asking them to stick their hand on it, while listening for a sizzling sound.
2. As a design limit in your future products to make sure you don't injure any customers.
3. It can be the lower operating limit for your new electronic branding iron. When the body-piercing trend wanes, I predict designer branding will take its place with the hip crowd. A machine that can brand the pattern of your choice into human skin (like a dot-matrix printer, only *crispy*) will make you a fortune.

T-TYPES FRIED MY BRAIN

I was attracted to an unmarked lab door by the unmistakable aroma of frying bacon. The smell drew me past the NO ADMITTANCE— CONFIDENTIAL signs, through a maze of lockers and racks of humming test equipment to a mysterious, unattended lab bench. On the bench, at the center of a web of dangling cables, was a shiny metal box, on top of which was a glass dome. It looked like an Electronic Cake Saver from the Sharper Image catalog. Soft fizzing and popping came out of the dome.

What next caught my eye was a device about the size of a clock radio. It was my thermocouple meter that had disappeared six weeks ago! I was about to snatch it and dash for the door, when I noticed something even stranger.

"Ugh," I said, shuddering at the sight of T-type thermocouple wire. I am a lifelong J-type thermocouple man, and here was the sin of T-type wire attached to my beloved J-type meter. I set the meter down gently near the cake saver gizmo. The bacon smell was coming from under the dome, as was the bundle of T-type wire.

"Hey, dude," a voice from behind startled me.

Composing my most innocent expression, I turned. "Hi, Herb. What's cooking?"

"It's the next phase of the HBU," he said

"Right, the Human Brain Unit. That telepathic-telephonic project is still going?"

"Absolutely, although you've got to keep this secret. UPCHUCK, the Union of Psychics, Clairvoyants, Homeopaths, UFOlogists, Chiropractors and Kirlianists, doesn't like the idea of putting just a brain in a box instead of using a whole human being."

"How can you keep a secret from a whole group of psychics? If they really can… "

Herbie lifted the glass dome, revealing a pink blob. The smoky meat smell got stronger. "This is our real problem right now," he said. "We're working on the brain/machine interface." He lifted up the blob with his bare hand, making a sound like sweaty thighs peeling away from a vinyl car seat.

"Is that a real brain?" I asked.

"Yeah. But not human, and not alive. It's from a pig, just for prototype purposes. You can get them by the barrel from the federal government, if you know the right people. Turns out a pig brain is just as psychic as a human one, but less distracted by thoughts of sex and food. Anyway, I'm trying to get the temperature of the brain grid array, what we call the BGA."

Herbie pointed to the circuit board under the brain. On it were several rows of square black components, each with a grid of tiny pins sticking up. Thermocouples were glued to some of these components.

"Those pins stick into the brain?" I asked, backing away from the bench instinctively.

"It's the connection between the network electronics and the neural pathways in the HBU. The brain rests comfortably on this bed of nails. The trouble is that the déjà vu canceler circuits have to run at 2 Gigahertz, so these chips under the bed of nails get a little warm. Our brain guy says the BGA has to be less than 60°C, or we start to cook the HBU. That's why I'm measuring their temperature."

"Where did you get these thermocouple wires?" I said.

Herbie said, "Oh, I scrounged them from an abandoned test fixture, down in the basement next to the mail room. I found the meter in a box of old circuit boards under my desk. Why do you ask?"

"In a minute. What results are you getting?" I said.

Herb put the brain back under the dome, then showed me the temperature readings. "It's weird," he said. "The hottest BGA is only 53°C. But even though they are under the limit, you can hear the brain sizzling a little bit. Do you have a handbook on the thermal properties of brains? Maybe our brain guy is a little off."

"I'll tell you whose brain is a little off," I said. "Take a look at these thermocouple wires. Each pair has a blue wire and a red wire. That is the color code for T-type. But you have them hooked up to a meter that is calibrated for J-type wire. All of your temperature readings are wrong."

"Are you sure? I thought that J and T stuff just meant the wires are for use in different environments, like underwater or outer space. I even tested them out before I started using them. At room temperature they matched the reading of the lab grade thermometer on the wall."

I explained, "Room temperature can't test a thermocouple calibration. Remember how a thermocouple works? When one end is hotter than the other, it generates a voltage. The voltage is proportional to the *difference* in temperature between the two ends. So when both ends of the thermocouple are at room temperature, there is no difference, so it is not generating any voltage. Zero. Nada. All thermocouples behave the same when both ends are at the same temperature, so you can't use a room temperature reading to tell them apart. You could even hook one up backward and get the right reading. Put a copper wire across the terminals of the meter and it would read the correct room temperature."

Herbie looked confused. "How does that work? Shouldn't the meter just read zero then?"

"The meter makes it easy for us. Inside is a sensor that measures the temperature of the meter itself. The meter takes the voltage from the thermocouple, converts it to a temperature *difference,* then adds it to the meter temperature. If the thermocouple reads zero volts, it adds zero degrees to the meter temperature and shows that on the display."

"OK," Herbie said, "but how far off could I be? Aren't the T- and J-types just a couple of degrees apart, like yards and meters?"

"Let's take a look." I pulled out my pocket edition of the ASTM *Manual on the Use of Thermocouples.*

Air temperature in Herbie's lab was 25°C. The BGA reading was 53°C, so the presumed temperature difference measured by the thermocouple was 28°C (see Figure 1-1). (Remember that thermocouples only measure the temperature difference between one end and the other—in this case, one end is the BGA and the other is the meter, which is at room temperature.) When a meter calibrated for J-type wire reads a 28°C temperature rise, that means it is detecting a voltage of about 1.4 mV. But Herbie used T-type wire to get that 1.4 mV. For T-type wire, a voltage of 1.4 mV implies a temperature difference

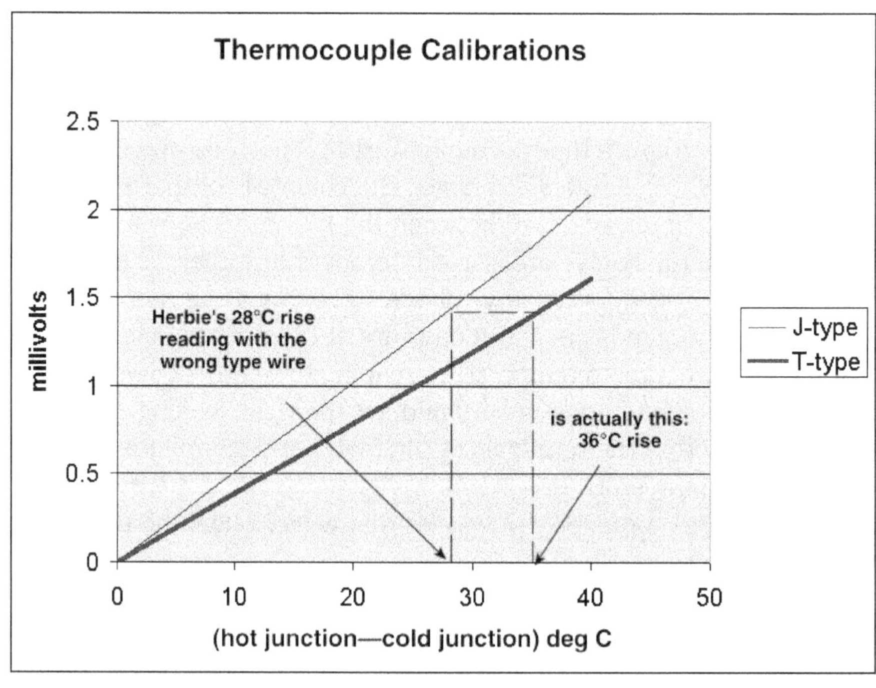

Figure 1-1

of about 36°C. Add that to the room temperature, and the real temperature of the BGA was 61°C.

"61 degrees!" Herbie moaned. "So we are over the limit. No wonder that brain is starting to fry. Now I'll have to add 8° to all my other test results."

"Not just 8°," I said, "the error gets bigger as the temperature difference increases. If you were in the 200° range, the error would be around 30°. Instead of trying to figure out all the errors, why don't you let me fix you up with some nice red and white J-type wire?"

Herbie shook his head. The brain was getting golden-brown around the edges. "This stinks. Thermocouple wire doesn't have any markings to say what type it is. How many kinds are there that I could mix up, and how the heck am I supposed to recognize them? And what about the meters? Where does it say J-type on this one? I made sure the cal sticker was up-to-date, and I figured that was good enough."

I picked up the meter. "You're right. It says J-type only on the back in small print, right next to the sticker with my name and phone number, which you didn't notice either. Table 1-3 gives you a hint how confusing thermocouples can be, so it pays to be careful when borrowing equipment."

There are about a dozen types available, but J, K and T are the ones you will see most often. Also, be on guard, because these colors are only standard in the U.S. and Canada. Europe has at least four different color codes. Japan's standard says all thermocouple types are white and red (no kidding!). And don't mistake thermistor or RTD temperature probes for thermocouples, even though they may look similar.

Table 1-3

Type	Color code	Metal pair
J	White/red	Iron/constantan
K	Yellow/red	Chromel/alumel
T	Blue/red	Copper/constantan
E	Purple/red	Chromel/alumel
R	Black/red	Platinum-rhodium/platinum

Meters can be tricky, too. Some are clearly marked with their type. Some can't be, since you can set jumpers inside to pick the type you want. Some require you to enter the calibration curve yourself via software (talk about opportunity for error!). The best (and for the same reason, the most dangerous) meters have built-in calibrations for all kinds of thermocouples, and allow you to select the type with a front panel control.

I haven't loaned that kind of meter ever since the time I caught Herbie's manager poking at one during a test. He was toggling through all the thermocouple types until one gave him the component temperature reading that he liked.

"How soon will you be needing the new thermocouples so you can continue your testing?" I asked.

He sat dejectedly, chin resting in his hands on the edge of the bench. "I don't know. This looks pretty bad. We may not continue the testing."

"I don't know if I'd call this a complete failure," I said, sniffing. "Got any eggs?"

PERMUTATIONS AND COMBINATIONS ADD UP TO JOB SECURITY

Herbie showed me this sketch of a new rack (Figure 1-2) that would interface to the human brain unit. "It's called the Psychic Port Module Exchange (PX) Rack," Herbie said. "It allows port cards for any of the four Telepathic Telecom Standards to plug into our Brainwave Telecom System."

"Four standards?" I said.

Herbie shrugged, "Yeah, ours and the three European standards. So you have four types of port card, and any card can go in any one of the 16 slots in any one of the six shelves in this rack, in any combination. And, oh yeah, they all have to work."

I looked over Herbie's drawings and started to make my Worried Face. "Let me get this straight," I said, "you have all six shelves stacked up with no baffles or gaps between, right? And there is a fan box on top, sucking out the hot air. So cold air goes into the bottom of the bottom shelf, flows through all six shelves and then shoots out of the top."

"Right," he said.

"I suppose the system has to keep working even if one of the six fans in the fan box goes dead," I complained.

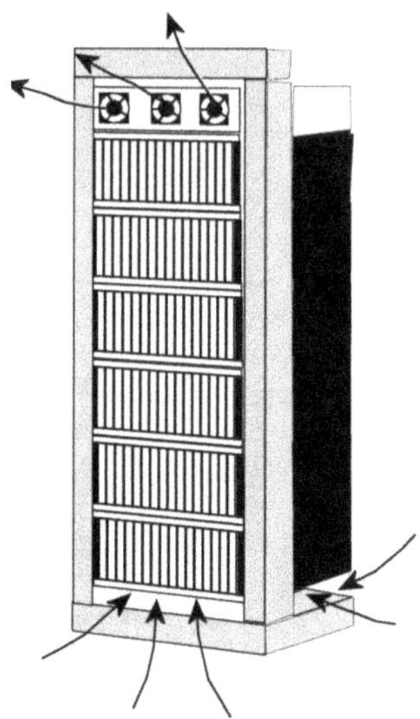

Figure 1-2 A simple rack cooling scheme. Air enters the bottom shelf, flows through all six shelves and then blows out of the fan box on top. Does this mechanical simplicity really need millions of thermal tests to fully validate it?

Herbie said, "At least until somebody can replace the dead fan."

"And you stacked up all six shelves like this with only one fan box—why?" I continued.

"This is the only rack configuration that would get us even close to the port density requirement from marketing," Herbie said. "Otherwise, our competition is going to crush us."

"Competition? I thought we invented this technology!" I said.

"It's very hard to keep trade secrets when telepaths are involved. Plus, the Patent Office insists that since you can't patent a natural phenomenon, they won't give us a patent for a *super*natural phenomenon, either. So anyone with a brain could come up with their own version," he said. "But don't get your undies in a bundle yet. This is just a proposed design. You tell us if it is thermally feasible. That's why I'm bugging you about it now."

"OK," I said, shaking my head gloomily, "looks to me like this will be a long job. Is TTM important?"

"TTM?"

"Time to market," I expanded.

"Obviously," Herbie said.

"Not so obvious from your rack design," I said, "It may be a thermally feasible design. But from a time to market point of view, it is a very, very bad design."

"Oh pshaw!" Herbie expostulated. "What difference does it make for time to market where we put the fan in the rack?"

"Here's the story," I said. "First of all, there are four different kinds of port cards so that means four times the amount of thermal analysis and temperature testing than if there were only one port card" (see Figure 1-3).

"Big deal," Herbie said, "four modules. There's only one shelf design, just repeated six times in the rack. It's not like there are six different shelves to check out."

"I have to make sure I'm testing each port card in its worst case thermal location in the rack, under its worst fan failure condition.

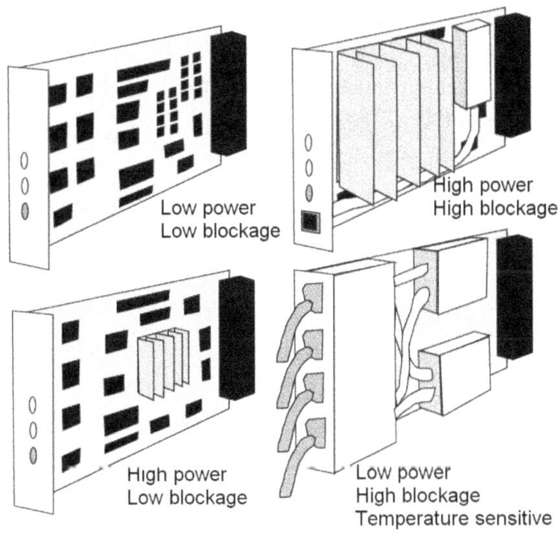

Figure 1-3 Four different port cards, which can be mixed and matched in any combination in the rack. Which card in which position will have the worst thermal problem?

Now, because you have stacked up the shelves into a nice chimney, I can't tell ahead of time where that worst case location is. Maybe it's in the top shelf, as you might guess. But maybe it is even worse for a high power board to be in the bottom shelf, if the slots above it are completely blocked by these other boards with their giant components.

"The temperature in any slot depends on where it is relative to the fans, the flow blockages above and below it, and the power dissipation of the modules below it. You've given me four port cards, some with high power, some with low power, some that block air flow a lot, and some that don't block air flow much. Plus one of these has components that can't take high temperature! And you can put any port card in any position in any combination. That starts multiplying out the number of tests I'll have to run to make sure I've found the worst case for each module."

"What a whiner!" Herbie said, "So you might have to run, what—four tests times six shelves—24 tests? At an hour each, that's only one day!"

"My department lets us go home at least every other night," I said. "But you've underestimated the number of tests a little. Let's see—four types of card that can stack in any order in six shelves, 16 slots per shelf, six different kinds of fan failure—that comes out to exactly 2,359,296 separate thermal tests." (I did use a calculator for that estimate.)

Herbie scoffed, "Two million! Don't exaggerate. It only took you 682 tests to do the recent thermal validation for Joe's new rack!"

I said, "That's right. But each shelf in Joe's rack has its own separate air inlet and outlet, and its own fan tray, so I was able to test each shelf independently. That cuts down the number of permutations and combinations to a reasonable level. The main reason it was even as high as 682 tests was that they had 53 kinds of power converter bricks they wanted to try out."

Herbie snatched my calculator and punched in the numbers himself, moving his lips as he ciphered. "Two million and change," he concluded. "But can't you eliminate some of the combinations right off the bat?"

I rubbed my chin and said, "Well, if we assume that the high power, high blockage module will be the hottest, so I don't need to measure any temperatures on other modules, that reduces the number of tests down to a mere 589,824. And if we assume we only need to test the four fan failures that are nearest any slot, instead of all six, it reduces to just 393,216 tests. That's still going to keep me busy for a while. What year did you say you wanted to release this design?"

"*This* year," Herbie said.

"There is also the problem of providing me a fully populated rack for testing," I continued. "Usually you give me a hassle when I ask you to provide just three working cards. This time I'll need a fully working rack, plus a full complement of each type of module—that's 384 boards. At $4,000 each, that would cost about $1.5 million."

Herbie nearly swallowed his bubble gum. "I suppose we could arrange to *borrow* one shelf's worth of cards, maybe," he said.

"Borrow?" I said, "I think not. We're going to need that fully-populated rack in my lab *permanently*. After all, every time you make a significant component change on any port card, we'll need a platform to test that change thermally before we can release it. We might have to re-do all 393,216 tests every time you want to try a cost-reduced voltage regulator. Oh boy, I think I'm going to have to put in a request for three test technicians and a bigger lab! Will your project manager be willing to fund them?"

"But how do our competitors do it?" Herbie asked, "They don't do two million thermal tests!"

"How do we know for sure what their port density is?" I said.

"Marketing downloaded an artist's rendering of a rack from their Web site, and counted the faceplates using a magnifying glass," he answered.

"How about this?" I proposed. "We toss a couple more fan trays and some baffles into our rack. I can handle a two-shelf stack, no problem. Just leave them out of our artist's rendering."

"But isn't all that testing job security for you?"

"Not if the company goes under while I'm still on test number 289,125," I said.

POWER CONFUSES, AND VARIABLE POWER CONFUSES ABSOLUTELY

Herbie summoned me to his lab. Around his bench skulked his cronies, Curly, Stretch and Odie. They tried to hide that it was a trap, with me the intended rat.

Herbie pointed to the center of a spiderweb of cables and scope probes: "Here's the bait—I mean, the board—I wanted you to see."

"It looks vaguely familiar," I said, peering at the prototype put together with standoffs, perf board, duct tape and Lego-brand blocks.

"Does *this* refresh your memory?" Herbie asked, producing one of my temperature prediction reports the way Perry Mason would flourish a motel registration card under a philandering witness's nose.

Recognizing the trademark color temperature map from the Therminator[1] thermal analysis software package, I said, "That's the psychic/optical port (POP) card that I said would never work. In the "**Conclusions,**" I even wrote: *This board will be so hot, the name should be changed to POP Tart.*'"

The cronies snickered. "This time I think I was smart *not* to listen to you," Herbie said. "We built it anyway. I just did a quick check of what were supposed to be the hottest and coldest components, according to your report. And I got exactly the *opposite* results from what you predicted. The coldest is the hottest and the hottest is the coldest. Explain that, Heat Sink Boy!"

I perused the bench more closely to see what Herbie had done wrong this time. Maybe he had plugged in his thermocouples backward again so that high temperatures read as lower than ambient. Herb's cronies gathered around to prevent my escape.

Herbie smiled wryly and said, "This time I have witnesses. You can't twist it all around in your next book and blame it on me."

"Yeah, no twisting," Odie echoed.

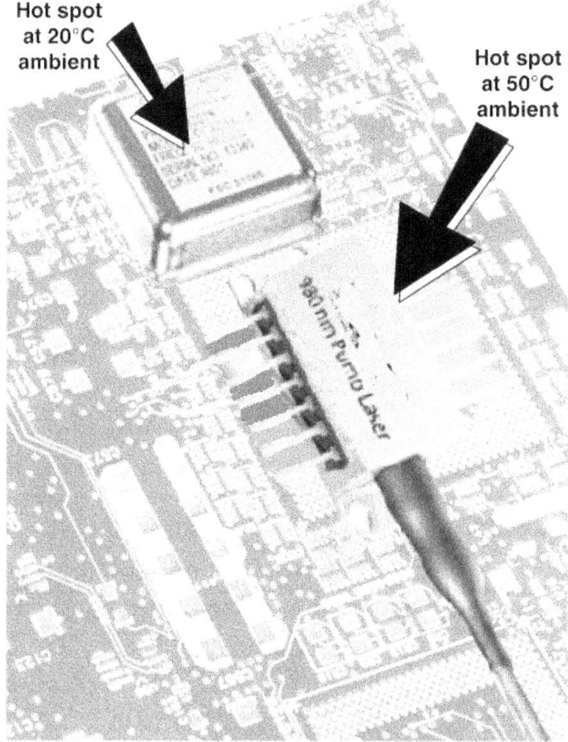

Figure 1-4 When power can change with temperature, the hot spot can move around as the ambient goes up.

"I did admit a mistake once," I said, "but that turned out to be wrong. So which one of these components is the hot spot and which one is the cold spot?"

"*You* said the laser transmitter would be the hot spot, and the 1 Megahertz crystal oscillator would be the cold spot. But according to the thermocouple probes I borrowed from you last week without asking, the laser is barely above room temperature and the oscillator is over 70°C! It's too hot to touch. Go ahead, put your finger on it and find out!" Herbie said.

I read the meter myself (see Figure 1-4). Room temperature was 23°C, the laser transmitter was 28°C and the oscillator was 74°C. To confirm that, I lightly brushed my fingers over the surface of the oscillator casing. It was very toasty.

"Ouch," I said with a flinch. "Looks like you're right, Herb."

"Yes!" Herbie exclaimed triumphantly, then led his cronies in a ceremonial dance around the lab bench.

I added calmly, "But that doesn't mean I was wrong."

Herb halted and the three guys did a chain-reaction crash into his back. "No twisting! If I'm right then you have to be wrong," he cried.

I explained, "But you're comparing my temperature prediction with measurements done under different conditions. My report gave component temperatures at worst case—that is, when the ambient air is 50°C. Your little demonstration here is at only 23°C. Quite different."

Herbie smiled confidently and said, "That won't work on me! I know that the component temperatures go up as the ambient goes up. You yourself taught me that if the ambient goes up 30°, the component goes up 30°, too."

"Yeah, 30°!" Odie chimed in.

Herbie silenced Odie with a glance, then continued: "And I didn't claim you got the wrong values for temperature. I said you goofed on which component was hottest and which the coldest. Even if the ambient changes, the hot spot should stay the hot spot, and the cold spot should stay the cold spot. So you still blew it."

"Everything you say is true, *unless...*" I said, and saw their smiles freeze half-formed, "...unless the power dissipation of the component changes with temperature. The temperature rise of a component is directly related to its power dissipation. If the power is constant, then that temperature rise is constant. If the ambient rises, the component rises with it, degree for degree.

"But suppose you put a cheap power resistor across a constant voltage. At 20°C ambient you measure its temperature to be 75°C. That's a rise of 55°. By your method, if the ambient goes up to 50°C, you'd predict the resistor would go up to 105°C.

"But in reality, the resistance of your cheap resistor goes up with temperature. Higher resistance draws less current from a fixed voltage, and less current means less power dissipation (remember I^2R?). Less power means lower temperature rise, so the actual resistor temperature at 50°C ambient is lower than 105°C."

Herbie began to look worried. His cronies were writing on their hands trying to remember Ohm's Law. "But this laser transmitter ain't no cheap resistor. It's got an expensive thermo-electric cooler inside!" he said.

"Exactly," I said, "I took the time to read its data sheet when I did the Therminator analysis. Not only is the power dissipation not constant with temperature, but the circuitry inside actively changes the power with temperature on purpose! The wavelength of the laser diode fluctuates with temperature, so the device was specially designed to maintain the diode at exactly 25°C, no matter what the ambient. The reason they included that thermo-electric cooler was to keep the output wavelength of the diode constant. One funny side effect is that the thermo-electric cooler can dissipate lots more heat than the laser diode it controls. When the ambient is about 20°C, the power dissipation is only about half a watt, which is the heat from the diode alone. But when the ambient goes up, the thermo-electric cooler kicks in to keep the diode at 25°. The hotter the ambient, the more power it takes. At its upper limit, the transmitter can be dissipating 5 watts!

"So when the ambient is normal, like now, the power is very low, and the case is only a few degrees hotter than ambient. At 50°C ambient (the condition I assumed when I predicted its temperature) the power will be 10 times higher, and the transmitter case will be the hottest thing on the board."

Herbie opened and closed his mouth like a fish, stifling his snappy retorts as his brain gradually accepted the concept. "But how come you were so wrong on the crystal oscillator, then?" he finally said.

I sighed, then said, "At normal room temperature, my prediction *appears* to be wrong, for a very similar reason. The power dissipation of the crystal oscillator changes with temperature, too, only in the opposite direction. The natural frequency of a crystal changes with temperature, so to keep this one from varying with the ambient, it has a thermostat and a heating element to regulate its temperature to around 75°C. It's an *oven* oscillator, because it keeps itself hot all the time. It's the same idea as the laser, but a simple heater is a lot cheaper than a thermo-electric cooler.

"When the ambient is 0°C, the heater can draw 10 watts trying to keep the crystal hot. At 20°C it might take 5 watts. But at 50°C, maybe it only takes half a watt. Right now on your bench we're measuring the crystal case temperature around 74°C, making it the hottest thing on the board. But at 50°C ambient, the case will probably still be only 74°C. But then it won't be the hottest thing on the board anymore, because everything else will have gone up 30° (or more, if you look at the laser transmitter). I took all that into account when I did my thermal simulation."

Herbie nodded thoughtfully. "So if I put this board in a chamber at 50°C, the oscillator will stay 74°C, but the laser transmitter will go up to 98°C, like it says in your report? Not 55°C, like it would seem from the 5° temperature rise it has on my bench now?"

"98°C, yup," I said, "and as I seem to recall, that's about 28° over its operating limit. That's why I told you it would never work."

Stretch whispered to Herbie, "Maybe you ought to catch your boss before he gets to the Big Guy's office. He was going to tell how you got the POP card to work."

Herbie choked and dashed down the hall. His cronies giggled. Lab loyalty, they say, is measured in nanoseconds.

Curly picked up the fallen thermal report. "That thermo-electric thing was cool and all, but how do they do this?"

He was pointing at the microprocessor on the color temperature map. It was clearly marked "4°C."

"Oops," I said, "that's a typo. Should be 94°. How did Herb miss that one?"

Notes

1. Therminator is a fictitious software tool, similar to several real software packages available today for analyzing air flow and temperature in electronic assemblies. It is based on computational fluid dynamics (CFD). If you can create a 3-D computer model of your electronic hardware, Therminator can solve the fluid dynamics and heat transfer equations for the model. It is most frequently used to make predictions of component temperatures on circuit boards, and the post-processing software creates quite impressive color maps of the temperature distribution. For more details, see Chapter 17 in *Hot Air Rises*.

How to Get Percent Error 100% Wrong

It began on a rainy Wednesday morning a few years ago. I was inno-cently clicking away at my computer, when suddenly a new window popped open all by itself. It displayed the seal of the Office of Metrics Enforcement, a Department of the Pie Chart and Bar Graph Division. That dissolved into a close-up of a man's face.

"Ah, Kordyban," his voice spoke from my computer's speaker. "I see you are at your desk. Please wait there for me. I will be in your office in 8.4 minutes."

Before I could even figure out how to respond, the window van-ished with an audible chirp.

Hans arrived right on time, toting a laptop and wearing two beep-ers and a cell phone on his belt. As soon as he sat down he started poking at some kind of handheld digital gizmo with a stylus.

"How are you doing, Hans?" I asked. I remembered him from the hot dog line at the company picnic.

"Getting 0.087% better every day," he said with a measured grin. "What is more important is exactly how *you* are doing these days."

"Oh, fine, I guess," I said hesitantly.

"My sources report that you have indicated, on more than one occasion, that your thermal simulation tool, Therminator, can predict component temperatures to within 5°," he said.

"Plus or minus 5°C. Yes, that sounds like something I would say," I admitted. "Before a product is built, I use the Therminator to predict component temperatures. After the first prototype is built, I measure the actual component temperatures. If I did a good job, the temperatures can match within that range."

"That has all the earmarks of a *metric*," Hans said. "You have a process, you have a measure of the quality of its result. We now assign to you a target in terms of this metric for improving your thermal simulation process. I have arbitrarily selected the goal of 10% per year, which should be achievable with an average effort."

"10% of what?" I asked, a shiver running up my back.

"That is up to you. I am now e-mailing you a form in which you define your metric. Each year you will provide data on how your process has improved," he said.

I glanced over at the e-mail scrolling up on my screen. There was a subtle beep, and when I looked back, Hans himself was gone.

Like everyone else, I was too busy that year to bother with formal process improvement. I needed to stall. So I defined my metric as the percent error between my temperature prediction and the final temperature measurement for only the hottest component in each product that I tested.

$$\% \text{ error} = (T_{measured} - T_{predicted})/T_{measured} \qquad \text{Eq. (1-1)}$$

The hottest components were usually about 100°C. With an average difference between prediction and measurement of 5°, that gave me a starting metric for error of 5%.

My brilliant idea for process improvement for that year was to change my units for temperature recording from Celsius to Fahrenheit. The difference between prediction and measurement became 9°F (since 5°C = 9°F), and the hottest components were now 212°F (see "**Summary of Lessons**" at the end for how this works). That gave me a new metric of 4.2% error. That was more than enough improvement to meet my 10% goal (10% of the initial error rate of 5% is 0.5%)—without changing anything!

By the next year, I was even busier and needed to stall again. I changed units again, this time to the Kelvin scale. My error rate was now 5°K out of 373°K, which was only 1.3%—a whopping 69% improvement over the previous year's results. Hans rewarded me with an electronic coupon good for 69% off an ice-cream sandwich in the cafeteria.

When the next year's reporting period rolled around, Hans appeared in my office. "Kordyban, what are you going to do now?" he droned. "You are out of temperature scales. Rankine will give the same percent error as Kelvin."

"I know that, but, hey!" I said. "You mean you knew I was playing numbers games all along?"

Hans didn't even blink. "From day one. But you followed our procedure to the letter so I could not fault you. But now that you have been assimilated into the system, you must cooperate and actually look at your thermal simulation process, to make some real changes."

I said, "I *have* been thinking about how to refine this metric. Really! I got to thinking about it during my last haircut. My buddy Joey and I go to the same barber every month. He's got hair like Grizzly Adams, and hair is something I don't have a lot of. When the barber's done, we each give him a nice tip, even though Joey's hair comes out looking great, and I end up still looking like me. I don't measure the barber's performance based only on the end results. I measure him on how good he did with what I gave him to start with."

Hans squinted at my bald pate, nodded, and said, "Continue."

"Even the best barber in the world can't make me look like James Dean. So it occurred to me that we ought to look at thermal simulation the same way. We should apply a metric only to the part of temperature prediction that it actually has some control over," I said.

"Elaborate," Hans said.

Here is my elaboration: Let's say we want the Therminator to calculate the temperature of a single component. It uses an equation like this:

$$T_{component} = T_{ambient} + Power/(h \times Area) \qquad \text{Eq. (1-2)}$$

My original metric rates the Therminator on how well it calculates $T_{component}$. But how much of $T_{component}$ is it responsible for?

Not for $T_{ambient}$, the air temperature of the room where the equipment operates! Ambient temperature is an input to the problem. For the typical problem, ambient is 50°C. For a component temperature of 100°C, the ambient makes up half its value. The Therminator can be held responsible for at most the other half. Let's toss out the ambient, then, and include only the portion of the component temperature computed by the Therminator:

$$T_{Therminator} = Power/(h \times Area) \qquad \text{Eq. (1-3)}$$

What about the component **Power**? Again, this is an input value to the problem, not calculated by the Therminator. Power is an educated guess made by the electrical design engineer. From Eq. (1-3) you can see that $T_{Therminator}$ is directly proportional to the Power. If there is a 10% error in the Power, there will be a 10% error in $T_{Therminator}$.

The same thing applies to **Area,** which is the surface area of the component. $T_{Therminator}$ is inversely proportional to this input number. An error in the **Area** could be just as important as an error in the **Power.** But that is much less likely to happen, because the geometry of components is not subject to guesswork.

That leaves *h,* the convective heat transfer coefficient. This is the only term in the equation that is not an input to the Therminator. Finding *h* is the core of the thermal simulation process, and the thing we should be measuring the error of.

Hans quivered with analysis. He activated his wearable computer and scanned his metrics database through his ocular implant. "According to your impromptu song and dance, component power is the key driver of temperature prediction. We have several reports on the accuracy of component power estimation. Our data shows that power is frequently estimated as much as 200% too high for new components. The typical error in power estimates is in the range of 25%," he said. "Your typical component temperature is approximately

70°C. With the typical ambient of 50°C, that gives a typical $T_{Therminator}$ of about 20°. Given that your typical prediction error is 5°C, the percent error is 25%.

"The input data to your process has a typical error of 25%, and your process output has a typical error of 25%. Any attempt to improve your thermal simulation process would appear futile."

I seemed to be off the hook. I offered, "Maybe the way to get better thermal simulation results is to get more accurate estimates of component power. But that is outside my control. I get that data from the hardware designers."

Hans said, "Give me a list of their names."

I put my fingers on my keyboard and began to picture some hardware designers in my head. Before I could start typing, I heard a subtle beep from Hans.

"Thank you," he said, "I will deal with them."

Summary of Lessons

In case this story was too bizarre and confusing, here are its basic ideas:

1. This is how you convert temperature scales:

$$°F = 32 + 1.8 \times °C \quad °K = 273 + °C \qquad \text{Eq. (1-4)}$$

2. When comparing two temperatures, to get a percent change or a percent error, the correct way is to compare the temperature rise above ambient, not the temperatures themselves; 100°C is *not* twice as hot as 50°C.
3. The biggest source of temperature prediction error is power estimate error.
4. When the temperature prediction is very close, I take the credit. When it is way off, it is not my fault.

Section 2

Fans: Increasing the Air Flow and the Trickiness of Your Cooling System

It would be better if we weren't surrounded by fans. Then we wouldn't take them for granted. They seem so simple—an electric motor spinning a paddle wheel to throw air around. If we weren't used to have that old clunky oscillating fan to cool us off on those hot summer nights, we wouldn't think they were so simple. If we had never heard of electric fans; if, instead, we first encountered them, in a Ph.D. seminar on Rotational Aerodynamic Gas Acceleration Systems, we wouldn't make the blunders we do with them in cooling electronics. We would respect fans for the mysteries that they are.

It is tempting to think we can solve thorny thermal problems by introducing a fan or two. In reality, we are trading one set of problems for another, as the stories in this section illustrate.

ELBOW ROOM

W hen surveyed on how my department could improve the timeliness of its services, one manager responded, "Clone Kordyban." The number of projects requesting thermal simulation work had been increasing faster than my capacity to crank them out, so a backlog had formed.

This suggestion was taken seriously (as seriously as any other employee suggestion), and after a long and diligent, cross-functional, team-oriented, company-sponsored development process, an offspring was produced (see Figure 2-1). Her code name is Fiona. Although DNA testing confirmed approximately 50% Kordyban genetic material, it became apparent that some training was going to be necessary to bring out her innate thermal analysis skills. The project feasibility was called into question, some experts projecting that the training may take between 21 and 25 years to complete.

The training effort was initiated with great gusto. Hopes were pinned on the new GOOEY interface in the latest release of the Therminator software. Instead of relying on mouse and keyboard input, the new interface was primarily drool-operated.

Figure 2-1

Ideally, it was expected that adding another me would double the output of thermal simulation. Fiona caught on quickly, but to date still has trouble with the k-ε turbulence model. No matter how good she got at running her own Therminator problems, we still tended to get in each other's way. For example, at 11 weeks old she still could not sit up on her own, and had to sit in my lap. What with all the goo-gooing at each other, and sharing of computer resources, our total output was somewhat less than the doubling we had hoped for.

Which reminds me of Herbie's less-than-ideal fan tray for the PSI-Cell System (the cellular base station version of the HBU). His fan tray was a simple metal box holding six fans that he planned to slide under each PSI-Cell shelf. He needed to know whether the fans he had selected would produce enough air flow to cool PSI-Cell's toasty circuit boards.

"I remember how a single fan works," Herbie told me. "The fan performance curve tells you the air flow rate versus the pressure. The more flow blockage in the shelf, the higher the pressure, so the flow rate goes down."

"OK," I said, glad that he remembered something.

"But how do I figure out the flow when I use more than one fan at a time?"

"According to all the textbooks, fan catalogs, the Therminator user's manual and even the TeleLeap GD (Good Design) guidelines, this is how fans add together in parallel. That is what you are doing, I think, since you are mounting your fans side by side at the inlet to your shelf. The other way to do it is called series, where you put one fan at the inlet and one at the outlet, but we'll talk about that some other time. This graph shows how the air flow rate of the individual fans adds together" (see Figure 2-2).

"At any particular value of pressure, you just add together the air flow rates of all the fans. For example, at zero pressure, one fan puts out 80 CFM (cubic feet per minute). Two will put out 160 CFM, and three give you 240 CFM," I explained.

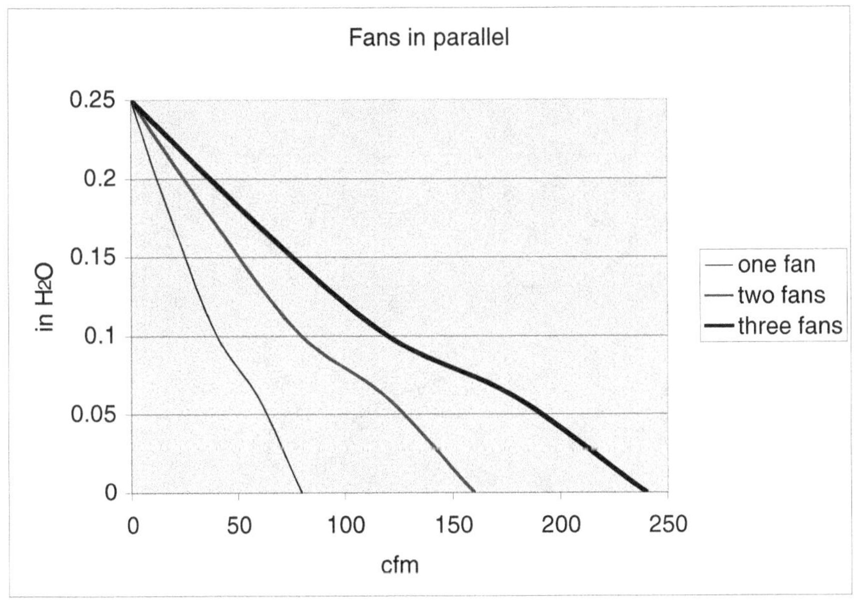

Figure 2-2

"That's a no-brainer, which is good for me," Herbie said. "I'm planning on using six of these fans. Added together, that gives me up to 480 CFM!"

"Don't forget that your shelf will block some of the flow, so there will be back-pressure. You won't be able to get the full 480 CFM, because the pressure won't be zero," I reminded him.

"That's OK," he said, waving me off. "I probably will need only 250 CFM, so even with some back-pressure my six fans should be able to do the job."

McCool Products built Herbie a prototype, and along with the sheet metal, wiring and fans, sent him the performance data that they had measured in their flow chamber.

Herbie came to me looking disappointed. "Can you explain this?" he said, showing me their graph (see Figure 2-3).

"What's the problem?"

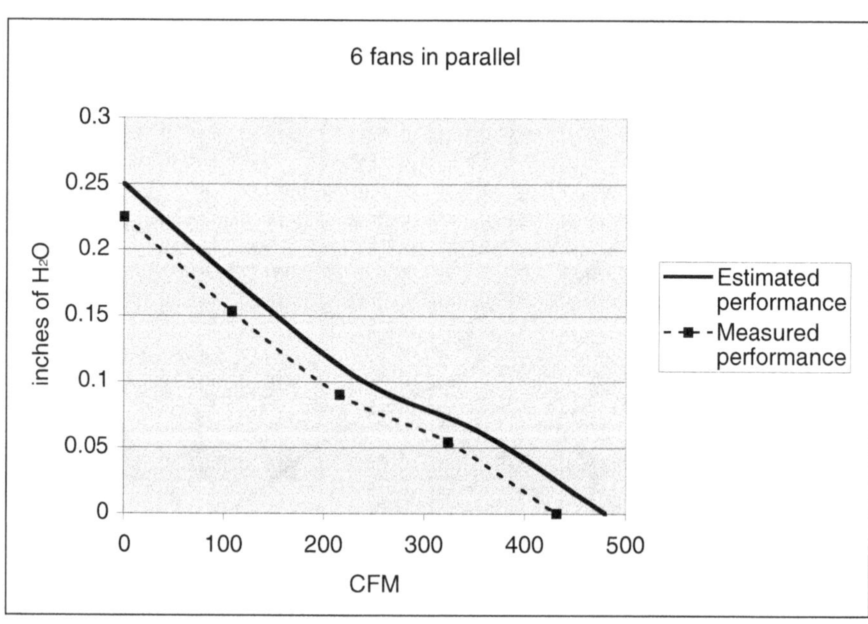

Figure 2-3

"This upper curve is what the fan tray was supposed to give me. I took the performance curve for one fan and added up the air flow rate at each value of pressure—just like you showed me. At zero pressure I should get 480 CFM. But this lower curve is what McCool measured for the fan tray!"

"Is this for only five fans?" I asked. "Or maybe a filter is cutting down the flow? Or there's a leak in the fan tray walls?"

"Nope. They said that this is what the fan tray does with no leaks, no filter, no restrictions on the inlet or outlet at all."

"Low voltage?" I tried. I honestly didn't know.

So I did a little digging, and began to find a lot of footnotes in fan vendor data. It seemed to be a widely known fact among fan experts, but it was little publicized to us customers. It seemed to be summarized best by this couplet from a 19th century English mineshaft ventilation manual *Fans O'erflowin'*:

> *If in parallel you place your fans too **near***
> *Their in- and outflows will tend to **interfere.***

Although the poem is widely known, and often parodied, the source (author, and his test data) is unknown. In fact, no one seems to know how close is too close, or how much the flow rate goes down.

But if you think about it a while, it becomes obvious that when you shove fans together, edge to edge, they'd have to interfere with each other. A fan by itself can suck in air from all directions. With neighbors, each fan has a more limited space from which to draw air. Also, the air swirling out of each fan collides with the air streams from the neighboring fans, in effect, restricting the outlet. On the inlet side, the fans are fighting each other for air to suck in, and on the exhaust side they are fighting for space to blow their air into. With all that tussling going on, there is wasted effort, so the fan tray gives a less than ideal performance.

"So how far do I need to spread the fans apart to get the best air flow?" Herbie asked.

"The last line of the poem is *very* helpful," I said, and quoted:

> *If 'tis with optimum spacing you think you should*
> *bother,*
> *'Tis the span whence fan no more interferes with*
> *another.*

The lesson is, with fans or thermal analyzers, that the sum of the parts doesn't quite add up to the whole, especially if you don't give them enough elbow room.

Breathing Room

In preparation for my meeting with Marketing, I grabbed a roll of duct tape and some other lab equipment. It was time for a demonstration.

A few days earlier, I had received a "carbon copy" of an e-mail between the Evil Twins of Marketing, Joel and Ethan. "When will the thermal study on the Quadruple Redundant Access Port (Q-RAP) system cabinet be done?" it asked matter-of-factly.

I had never heard of the Q-RAP system before, so I went to the one place I knew I could find out about it—not in the official Product Definition Document, but in the pile of unclaimed PowerPoint printouts in the copier room near the Marketing enclave.

Slides in hand, I searched every coffee machine alcove for Herbie. His initials were all over them.

"Oh, yeah," Herbie said, nodding over the slides, "I remember this. They took a perfectly good RAP shelf, and screwed it up by cramming six of them into a European-style cabinet so they could sell it to the Transcarpathian Telecom Commissariat."

"You went along with this?" I asked.

"I got stuck with routing the cabling. Other than that, Ethan and Joel did the whole design themselves. Engineering was cut out of the loop."

"Can they do that?" I said.

Herbie shrugged. "Seems like they can do anything they want, as long as they can get a customer to buy enough of them. When we developed the RAP shelf a couple of years back, your thermal test said that we needed to put a baffle between every pair of shelves. The boards were so hot in natural convection that we couldn't allow the hot air from the lower shelves to flow into the shelves above. That meant that we could fit only four shelves in a 7-foot-tall rack. Now Ethan claims that one of our competitors, TinCanTech, offers six shelves in a 6-foot-tall cabinet in Europe, so we have to be able to do the same thing. I said no way, based on the thermal tests. Joel just said yes way, we can solve that problem with fans, and that it was all your idea."

"My idea?" I said.

"Yeah," Herbie continued, "remember last Christmas, that big program review meeting?"

"Not especially," I said.

"Joel put up this slide showing six RAP shelves in a cabinet. You rolled your eyes and said sarcastically, '*Oh sure. If you want to do that you'll have to add a fan tray!*' Well, buddy, they took that as a design recommendation," Herbie said.

"Sounds to me like somebody needs to be educated, old-school style," I said. So Ethan and Joel were invited to a thermal design review.

"It's about time," Ethan said, "A little high-intensity face time will put this thermal milestone issue to bed."

"Absolutely," Joel echoed, "We want to hear your swag at this thermal thing, then put your John Hancock down on the bottom line so we can broom this out the door, so we can meet our customer's expectations of global high-quality delivery objectives."

"Whatever," I said. "To get started, I have one small issue with the Q-RAP cabinet design: the amount of space assigned to the fan tray" (see Figure 2-4).

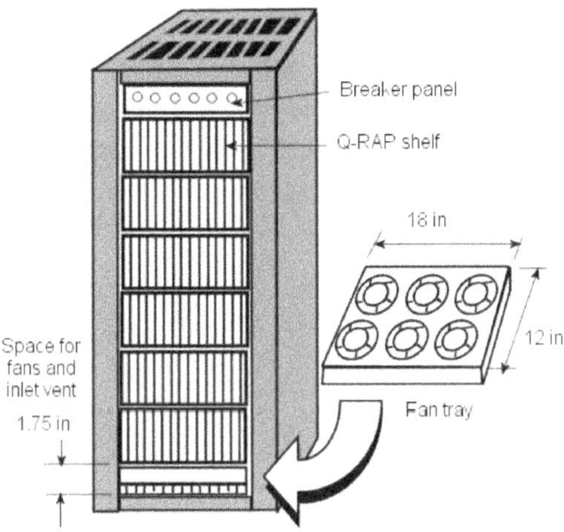

Figure 2-4 The Q-RAP Cabinet does not leave much breathing room.

"We do, too," Ethan said. "We know that the fans are only an inch thick, but we allowed a full 1.75 inches for the fan tray. We would have liked to have used that extra $^3/_4$ inch of cabinet height to cram in some more RAP modules, but they just don't fit. So you can have all that extra space for your precious fans."

"Ah, that's where we differ, gentlemen," I said. "Whereas you think you have been generous in allotting space to fans, I, on the other hand, think you have been a trifle stingy. I don't need any tests to know that your cabinet will have pretty skimpy air flow. The 1.75 inches for the fans includes the space for the inlet vent!

"The fans are an inch thick by themselves. Then you need an air filter, to keep from coating the RAP electronics with dust. The minimum air filter thickness that will work is $^1/_2$ inch. That leaves only $^1/_4$ inch, assuming you can rest the air filter right on top of the fan blades (which you *can't*), for the actual inlet area."

I motioned for Herbie to begin the demonstration. As I continued to speak, Herbie stuck a cocktail straw into Ethan's mouth, then sealed up his mouth and nose around it with duct tape. It wasn't a large bore plastic milkshake straw, but one of those annoying red straws about four inches long you get in a mixed drink, about the diameter of a toothpick, that you don't know whether to drink through or stir with.

I explained, "This will give you an idea what a cooling fan feels like. Joel, you are a fan tray with an unrestricted inlet. Breathing OK?"

Joel sniffed in a lungful of air and nodded.

"And Ethan, you're the Q-RAP fan tray with a $^1/_4$-inch inlet tall inlet vent. Doing OK?" I said.

"*Umm-mmm*," he said, still looking chipper. His breath whistled slightly through the straw.

"Here's the lesson about fans," I said, "to get the most air flow, you shouldn't restrict their inlet or outlet. That means the inlet vent that leads up to the fan tray should have about the same area as the fans themselves. Your fan tray is about 18 inches wide and 12 inches deep. So, ideally you should have a vent 18 inches wide by 12 inches deep, like this" (see Figure 2-5).

"But isn't air pretty flexible about going through holes and around corners and stuff?" Joel asked.

"What do you say, Ethan?" I asked. Ethan was starting to perspire, but he still nodded that everything was OK. I admired his loyalty.

I said, "It is true that air flows around obstacles better than molasses, for example. So you can cheat a little from the ideal and

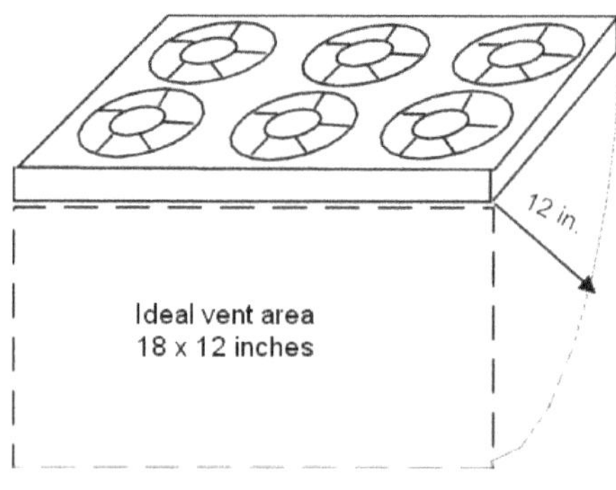

Figure 2-5

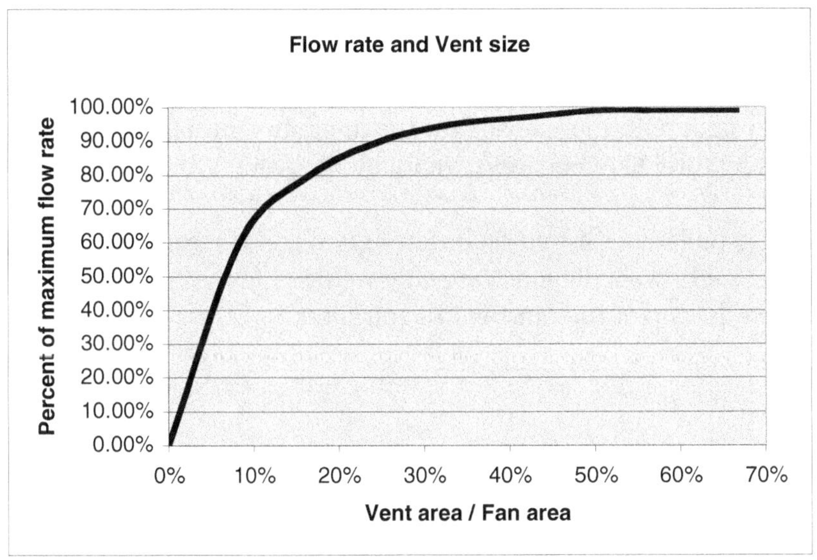

Figure 2-6

still not lose a lot of air flow. Our experience with other fans showed that you could reduce the inlet vent size from 100% to about 60% before you start to measure a significant drop in air flow. But after that, the flow rate starts to drop pretty fast, as you can see in this graph. With the teensy inlet vent that you have allowed, I don't think you will get even 20% of the air flow you expect" (see Figure 2-6).

Joel lost his usual elfin grin. "Are you saying that the space for the fan is going to have to be bigger? Like around 3 or 4 or even 7 inches taller? No way! We'd have to toss out one of the RAP shelves!"

Ethan's face changed colors, from red to purple. The whistling grew louder and faster as he went into oxygen deficit. But he, too, indicated with hand gestures that increasing the vent area was unacceptable. I could almost make out his argument as he grunted at the slides. Then he collapsed under the conference table.

"Should I take the tape off?" Herbie suggested.

"Is he dead yet?" I said, nonchalantly.

"Nope."

I turned back to Joel. "That is what is likely to happen to your Q-RAP cabinet, unless you allocate more space for the inlet vent."

Joel studied his partner, looked at the ceiling tiles for a minute or so, then said, "OK, more space for the inlet air vent. Maybe we can call it a feature, like, enhanced ventilation assembly, and charge extra."

Herbie pulled out a wad of air filter foam and the roll of duct tape. "Now," I said, "we'll demonstrate why we need to have 3 inches between the top of the fans and the air filter."

THE PATH OF LEAST RESISTANCE

L unch was promised, so I showed up for the "Howdy! Get acquainted" meeting with the folks from the TeleLeap Merger of the Month, Consolidated General International Corporation, of Vermilion Neck, West Virginia. Their whole R&D staff of 11 people were there.

The VP of Product Concretization explained what they were working on with a flurry of color slides. "It's like the old Iridium project, which was a worldwide satellite system for personal cell phones, except ours is targeted toward a slightly less upscale demographic. In the more remote mountain regions of this country, you will see plenty of our potential customers. Often they are not eagerly served by traditional telecom service providers. It costs a fortune to run telephone wires into these areas. But every trailer home with a pickup truck on blocks out front has a satellite dish! So we're developing equipment that will allow customers to connect to the global phone network directly through their existing satellite dishes. It's called Party Line 2000, a high-tech version of the old party line phone. To make it compatible with 1970s dish technology, we had to make it so any subscriber can listen in on anyone else's calls within a 60-mile radius. Marketing calls this a 'sellable feature.'"

Later, the big meeting broke up into small working sessions. I was grabbed by Zeke, the VP of Physical Design and head CAD operator. "We'd sure like you to give our new cooling system design the once-over-lightly," he said. "Our company was too small to do thermal analysis. We did things the old-fashioned way—built our prototypes, tested them and slapped heat sinks on anything that got too hot."

The PL2000 was a redesign of a satellite communication system for the Bolivian Navy. "We didn't use fans for that old design," Zeke said. "It was a shelf with 16 circuit boards in it, each one put out about 25 watts. All we had to do to make it work in natural convection was to stick baffles in between all the shelves, so that the hot air from the bottom ones didn't get into the upper ones in the rack. When we measured the hottest components, they were just barely acceptable.

"We've added a couple of extra doodads and features to the Bolivian job to make the PL2000. That jacked up the heat dissipation to about 30 watts per board. Our seat-of-the-pants design process tells us that if the thing barely worked at 25 watts, it ain't going to make it at 30 watts. Am I right?"

"Sounds logical to me," I said.

"But we also figured that if it was just a little bit over the limit of what natural convection could do, then just about any kind of fan cooling could easily do the job," he continued. "Even your high-falutin' GD guidelines say that fan cooling is about 10 times more effective than natural convection. So we didn't have to get too fancy. Just get the air moving and we should be all right, right?"

He showed me the system design they had come up with, with its fan tray at the bottom of the two-shelf stack-up (see Figure 2-7).

"The catalog that this here fan tray comes out of says it puts out 300 CFM, which sounds like an awful lot, which is good. But to get enough space for a fan tray in the rack, we decided that we would stack two shelves together on top of the fan. Heck, there's plenty more air than a single shelf can use! Then we heard that you had this amazing thermal simulation tool that might could actually predict what kind of air flow and temperature rise we might get out of this thing before we go ahead and build one. So before we plunk down

Figure 2-7 The PL2000: A fan failure just waiting to happen.

the money for the sheet metal and such, we thought you ought to have a crack at it," Zeke said.

"You betcha," I said. "If you give me a few dimensions for the shelf, the spec sheet for the fans, and a few more details about the boards that go in the system, we can get the Therminator to calculate the air flow and component temperatures."

Zeke grinned. "Let's do 'er!" he said, and we shook hands as if he had just sold me a hog.

Perhaps he had, I thought to myself later that week, as I began to build the thermal model of the PL 2000 in the Therminator.

With all fans running, the air flow through the shelves was pretty good, and the temperature rise predicted by the Therminator did seem to be OK. Then I ran the problem again, this time with one of the three fans disabled, to see what would happen during a fan failure (see Figure 2-8). Suddenly, fan cooling didn't seem to be the cure-all that Zeke had originally thought it would be

"What the heck is this?" Zeke screeched when I showed him the above results. "Why is the air flowing backward over here?"

"Well, sir, this fan here on the right side is failed…" I explained.

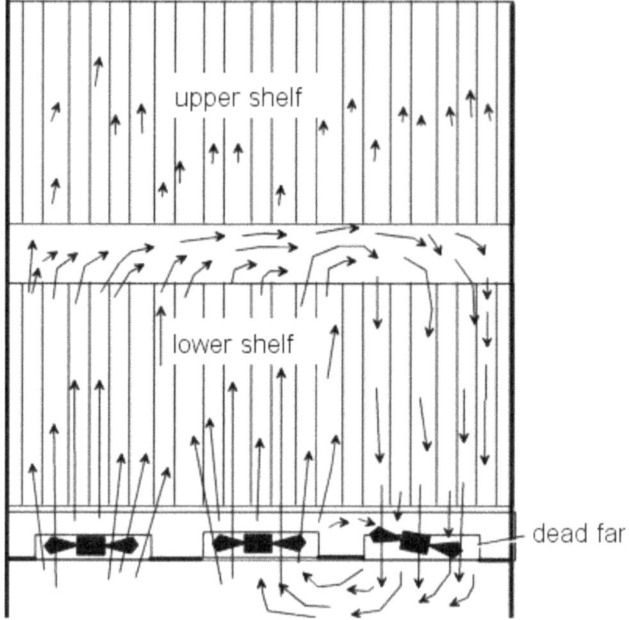

Figure 2-8 A dead fan quells no failures.

"Failed? Who said the fan's supposed to fail? Our plan is for them all to be working," he said.

"Fans fail. The motors burn out or the bearings seize up once in a while, just like lightbulbs popping. The only thing they guarantee about a fan is that they don't outlive the electronics. I assumed that the PL2000 is a high reliability system, that it is supposed to keep running, even when it sustains a single fault, like a failed fan," I said.

"Aw shoot, I suppose that's true," he said. "We can't shut off the shelf just because a fan gets stuck. I just never thought of that. Shoot. We never used fans before, so I never thought about what happens when one quits."

I said, "Sometimes redundancy doesn't work the way you expect. You'd think with three fans you have plenty of redundancy. After all, you barely needed any fans at all, so even if you lose one out of three, you should have plenty of air flow to keep the electronics running. Except that a dead fan is more than just a fan that stops blowing air. It becomes an air leak, the unofficial name for an unintended air flow path. This picture shows that air from the two good fans flows up through the first shelf, and then, instead of going up through the second shelf, it makes a U-turn through this space between the shelves, goes backward through the right side of the bottom shelf and blows out through the dead fan."

Zeke tried to look for a silver lining. "At least there's plenty of air in those slots, even if it is going backward."

"To make things worse," I said, "the air that leaks out of the dead fan gets sucked back into the intake of the good fans. So that hot air gets recirculated through the shelf over and over. The worst air gets over 120°C. That's not good."

"And it looks like the top shelf here isn't getting much air flow at all. The flow in some of these upper slots isn't much better than you'd get with natural convection," Zeke said.

"That's right," I said. "Overall, with three fans, this system works much worse than natural convection."

"That really eats my lunch!" Zeke exclaimed. "You make it sound like a fellow has to know what he's doing when he starts fooling around with fans."

"They can be tricky. And we haven't even started to talk about fan alarms, air filters, maintenance strategies or performance at high altitude. How tall are those mountains where you want to install this?"

Zeke rubbed his ample chin and scratched at his sideburns. "We been talking about branching out to the Rockies, too. Leadville, Colorado, is up there around 10,000 feet," he said. "But let's talk some more about how come that air flow gets to going backward through the shelf. Something don't seem quite right about it to me. Why does the air flow on the right side turn around and go back down through the dead fan instead of just going straight out the top of the shelf? Doesn't the air always follow the path of least resistance? And ain't there more resistance going back down through the shelf and the fans?"

I tried, "I suppose I could ask you to just believe it because the Therminator says so? How about a color printout of the velocity distribution?"

Zeke replied, "From what I hear about CFD, half the time you don't believe the Therminator results yourself."

"*Touché,*" I admitted. "Now about your theory of the path of least resistance. I think that's sociology, not fluid dynamics, but let's try a little experiment to see if it's right."

We rounded up a couple of drinking straws, and a piece of foam air filter. I stuffed the foam into the end of one of the straws, then laid them side by side on my desk. Then I invited Zeke to place his ear nearby and listen carefully.

"The empty straw has a much lower resistance to air flow than the one with the filter. Can you tell which one has more air flow?" I asked.

Zeke, who had been hunched over the desk very intently, looked at me suspiciously. "Are you pulling my leg? There's no flow through either one of them!"

"Are you sure? This empty straw just has to be the path of least resistance!" I said.

"But there's nothing making the air go," he said.

"Bingo!" I said. "Do you mean there's another important factor besides resistance to flow?"

A smile slowly spread across Zeke's face. "Something has to push the air. It doesn't just go by itself."

"Exactly. It takes a difference in pressure to makes air flow. Air moves from high pressure to low pressure, just like electric current goes from high voltage to low voltage," I said.

I had Zeke put both straws in his mouth and blow through them (see Figure 2-9). Lots of air came out of the empty straw. But a tiny bit still came out through the filter.

"By blowing into the straws, you are creating a pressure difference in them. Inside your mouth is high pressure, or at least higher than the air pressure in the room. The other end of the straw is at low pressure, so air flows out. But the air doesn't take the *path of least resistance.* Some, maybe even most of it, flows through the empty straw, but some still flows through the filter. The flow rate of each straw depends on the ratio of the two resistances," I explained.

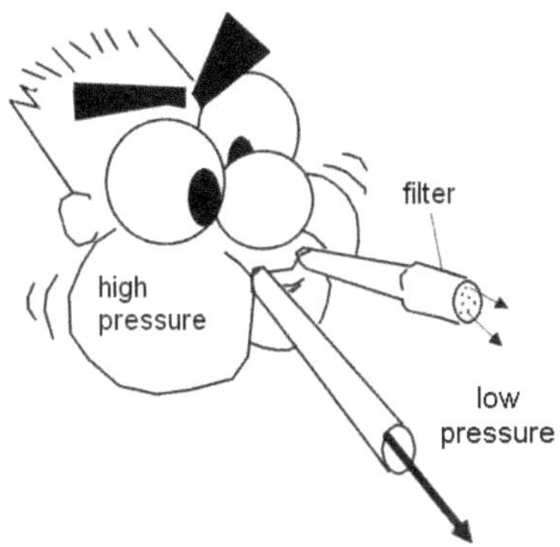

Figure 2-9

Flow follows the path of *most* resistance, too. Just not as much.

While Zeke continued to play with the straws, I sketched this diagram of a fan (Figure 2-10):

"This is how a fan works," I said. "When it spins, it creates high pressure on the exhaust side, and low pressure on the inlet side. Let's color in the low pressure areas with blue and high pressure with red."

Figure 2-10

Zeke held up his hands, "Whoa, amigo, I just happen to be color blind—a family trait common in my neck of the woods. Them color plots from CFD don't make no sense to me."

"OK," I said, "Then let's mark the places with pressure lower than room air with little minuses, and where the pressure is higher than room air, I'll mark with little pluses. At the inlet of the fan, the pressure is actually lower than room pressure. That is why air from the room flows into the fan. On the other side, the pressure is higher than the room, so air flows away from the fan."

"So minus indicates a vacuum?" he asked.

"Yep. Here's the next step. Let's mark our little pluses and minuses on the PL2000 system drawing, the one with the failed fan. I'll put pluses on the exhaust side and minuses on the inlet side of the working fans. It should look something like this," I said (see Figure 2-11).

"Look at the plenum, the space between the two shelves, the place where, if you were an air molecule, you have to decide whether to go up through the top shelf and out into the room or turn around and go

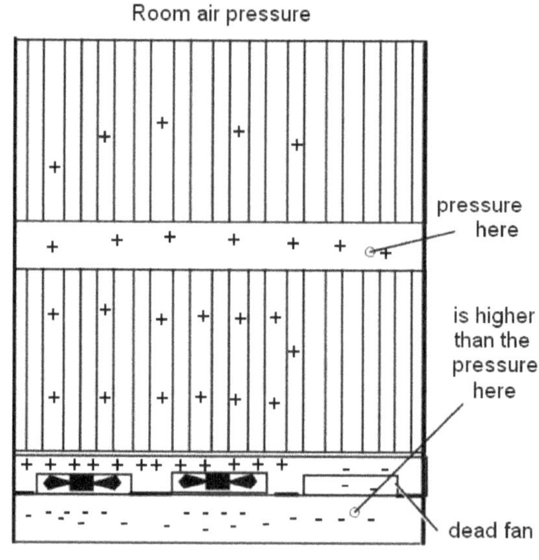

Figure 2-11

back down into the bottom shelf. The pressure in the plenum is +, higher than room pressure. But down here in the space under the fan tray, the pressure is −, lower than room pressure. For these slots on the right side, the pressure on top is higher than the pressure underneath, because the dead fan is a big short circuit in the air flow. So which way does the air in those slots flow?" I said.

Zeke hesitated, as if I had asked him to choose his card in a Three Card Monte game. "Uh, but the resistance…"

"Which way does air flow? From low to high, or high to low pressure?" I hinted.

"From high pressure to low pressure," he finally conceded. "So the air flows downward in these slots on the right."

Zeke kept studying the drawing, staring at the little pluses and minuses. "I'm still a mite unsure I get it. How do you know where the pluses and minuses go?"

"There is high pressure on the exhaust side of the working fans, and low pressure on the inlet side. After that it gets complicated. The local pressure at each point depends on the flow, and the flow depends on the pressure. At each spot there are multiple paths for air to take, each with different resistances and different pressure gradients. If you change the geometry a little bit, such as by increasing the open space above the fans, you might get a totally different flow pattern. That's why I use a computer program like the Therminator to keep track of all the flow and pressure and resistance bookkeeping," I said.

Zeke shrugged, "So I still have to just take the Therminator results on faith."

"That would make my job a lot easier, but I don't recommend it. Keep asking questions. It's the only way to keep me and the Therminator honest," I said. "One thing I do know is that the Therminator is a whole lot more accurate than an old spouses' tales, like *the air always takes the path of least resistance.*"

INCOMPRE-HENSIBLE FLOW

A few years ago, as an educational service, I created an electronics cooling "hotline." It was a toll-free phone number for anybody to call in questions about how to keep circuit temperature under control. After a few months I discontinued the service, mostly because the only questions I got were like this:

"Why are chili peppers considered 'hot,' and what temperature are they? They do make you sweat sometimes. And how come there isn't any 'cold' food?"

The next most common question was on the topic of confusion between volumetric flow rate and velocity. I kept getting this question over and over, and unless the guy was disguising his voice, it came from many different people. Here is an example of this question:

"I want to use a particular processor on my board, and its data sheet says that it needs 500 LFM of air flow. We have a 300 CFM fan. Is that going to be enough? How do you convert CFMs into LFMs? My boss thinks it's based on Roman numerals. You know, L is 50 and C is 100. Can you confirm this?"

As great as the Romans were at engineering, I think their influence on units of measure peaked with the cubit. Here's the short answer:

LFM is **L**inear **F**eet per **M**inute
CFM is **C**ubic **F**eet per **M**inute

They aren't the same thing, but they are related. It's like comparing *how fast* traffic is moving on the freeway with *how much* traffic is moving down the freeway. Obviously, how much traffic is moving depends on how fast the cars are going, but it also depends on how many lanes the freeway has, among other things.

Here is an example to illustrate how velocity, measured in linear feet per minute, is related to volumetric flow rate, measured in cubic feet per minute, for air flowing through a duct. But air is hard to visualize, so for the purpose of this example, let's say you are the quality inspector at the cheese factory, (note to my European friends: yes, in America cheese is made in factories by large industrial machines) and you are watching melted processed cheese-food spread (PCFS) flow through a big glass duct (see Figure 2-12). At the beginning of your inspection run, the duct is empty. You press a button to release the cheese flow and start your stopwatch at the same time. The

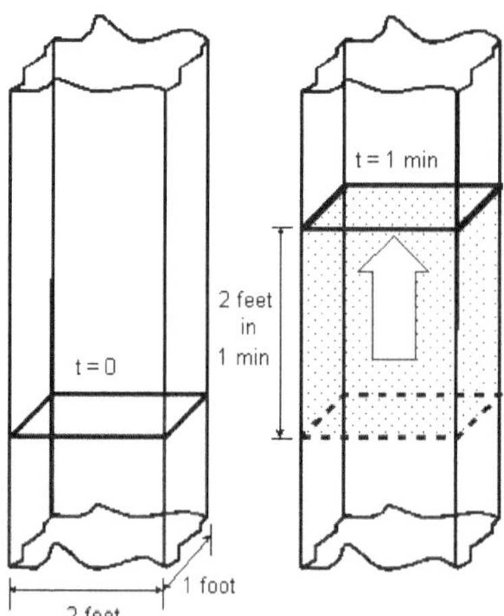

Figure 2-12

golden, gooey, yummy fluid immediately starts to flow past your inspection point.

After exactly one minute, you measure with a laser range finder (or a wooden yardstick from the hardware store) how far the front face of the PCFS has moved. You find that it oozed exactly 2 feet.

So the cheese velocity in your duct is:

$$\textbf{2 feet/1 minute = 2 feet per minute = 2 LFM} \qquad \text{Eq. (2-1)}$$

A "linear" foot is just a foot. People started to say linear feet, even though it is redundant, just to distinguish it from cubic feet, which is a measure of volume.

Now, how much cheese has flowed past? The cross-section of your duct is 2 feet wide by 1 foot deep, according to the sketch. The volume that has filled up with cheese during your 1 minute test is

$$\textbf{2 ft wide} \times \textbf{1 ft deep} \times \textbf{2 ft tall = 4 cubic feet.} \qquad \text{Eq. (2-2)}$$

That 4 cubic feet of cheese passed by your inspection point in 1 minute, so the flow rate of cheese was 4 cubic feet per minute, or 4 CFM.

Maybe you noticed that if you divide the volumetric flow rate (4 CFM) by the velocity (2 LFM) you get 2 square feet, which looks suspiciously the same as the cross-sectional area of the duct (see Figure 2-13). This is not a coincidence! The velocity (V) and the flow rate (G) are directly related by the cross-sectional area (A_{cs})

$$\textbf{G = V} \times \textbf{A}_{\textbf{cs}} \qquad \text{Eq. (2-3)}$$

If they are so closely related, why do people use CFM and LFM instead of just one or the other? Because they are not really interchangeable.

What happens later on down the PCFS line when the shape of the duct changes?

To save on expensive, cheese-proof glass after the inspection point, the size of the cheese duct is reduced, so now it is only 1 foot

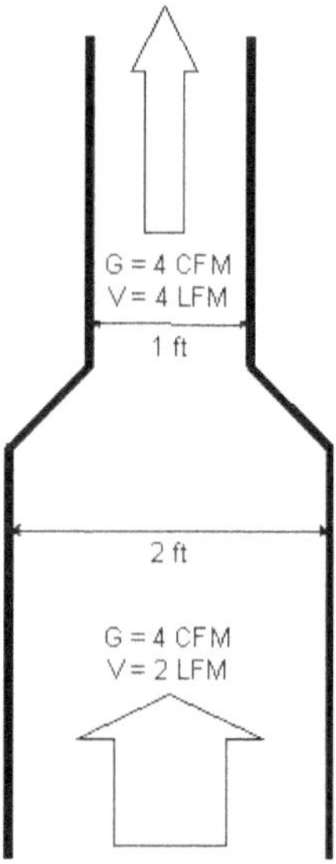

Figure 2-13

wide and 1 foot deep. What happens to the flow of cheese in the
smaller duct? There's no place for the cheese to hide, so all the
cheese that is flowing out of the big duct has to be flowing into the
smaller one. So the flow rate (G) has to still be the 4 CFM you measured earlier. But you are shoving all that volume through a smaller
duct, so something has to give. It does: the cheese velocity speeds
up. You can figure it out by rearranging the equation above:

$$\mathbf{V = G/A_{cs}}$$
Eq. (2-4)

The area of the smaller section of duct is 1 foot × 1 foot, which is 1
square foot. The cheese velocity is now double, or 4 LFM.

The flow rate of cheese hasn't changed, but the velocity has doubled, even though we are talking about the same cheese flowing
through the same ductwork. That is why you need to know both the
CFM and the LFM. The CFM though a duct is constant, but the LFM
changes with the cross-section.

Another reason for using both CFM and LFM is that flow rate is
important to some components, and velocity is important to others.

Fans, for example, are machines that make volumetric flow. A fan spits out the same volume of air per minute whether you hook it up to a skinny duct or a wide duct, so they are rated in CFM.

On the other hand, the temperature of a component, like the processor in the hotline question, depends not on the volumetric flow rate, but on the actual velocity of the air very close to the component. High velocity not only brings more fresh air in contact with the component every minute, but it creates turbulence, which improves heat transfer by better mixing the pockets of hot and cold air. So the component manufacturer writes the operating requirements for their part in LFM. They don't care about the total volume of air flowing through the chassis, or even through the single board slot. They only care about the air flowing right next to their part.

So how will you know the local air velocity for the processor, given a 300 CFM fan? (Let's pretend, first, that you know for sure the fan will put out 300 CFM, and not something a lot less because of the system back pressure. See Chapter 11 in *Hot Air Rises and Heat Sinks* for how to estimate the flow rate of a far.) All you need to do is figure out the cross-sectional area of the duct that the air is flowing through at the spot where your component is mounted (see Figure 2-14).

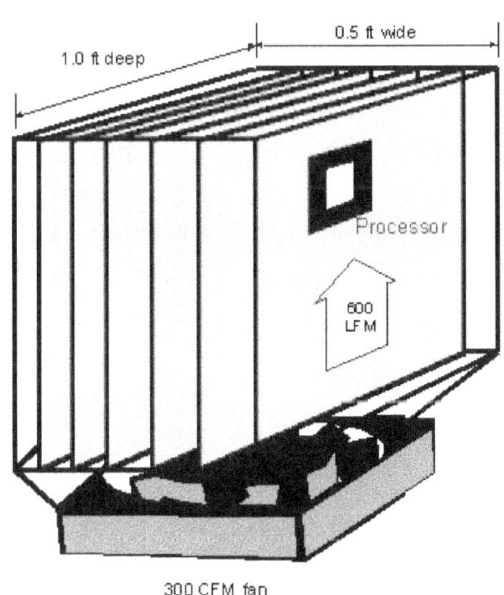

Figure 2-14

Let's say your fan blows into a box with five boards. The box is 1 foot deep and 0.5 feet across. If we neglect the thickness of the boards themselves (you don't have to if you want to be more accurate), then the cross-sectional area where the processor is mounted is 0.5 square feet. The velocity is given by

$$V = G/A_{cs} \qquad \text{Eq. (2-5)}$$

$$V = 300 \text{ CFM}/0.5 \text{ ft}^2 = 600 \text{ LFM} \qquad \text{Eq. (2-6)}$$

In this example, there is plenty of air velocity to cool the processor. But what if the box is 2 feet wide instead of 1 foot? The flow rate is the same (300 CFM), but the velocity at the processor would be only 150 LFM. Do you see why the data sheet specifies the velocity and not the flow rate?

Herbie might object that processed cheese-food spread and air don't act much alike. Air is compressible and cheese isn't. But cheese and air have more in common than you might imagine, especially the smell in some of the cubicles around the office. Air is compressible, which means when you apply pressure to it, it shrinks in volume, so how can I say the volumetric flow rate in a duct is constant for air? The pressure generated by fans is pretty darn small. It takes 1 atmosphere of pressure to squeeze a volume of air by 50%, but a typical fan used to cool electronics can generate only about 0.0008 atmospheres. The air volume might compress as much as 0.08%. So as far as you and I are concerned, air flowing through a shelf full of boards and cheese flowing through a duct behave a lot alike.

Now, back to that other hotline question about "hot" peppers and temperature. What if we melted some pepper jack cheese…?

CHAPTER 2.5

FAULT-TOLERANT COOLING

As a kid, I was always pestering my mother with questions like, "How come we don't have ice cream every night like they have at Jeffrey's house?"

Mom did not believe in talking down to kids. "We have different priorities at our house," she would explain. For a while that answer worked, because I thought the word "priorities" had something to do with my family being Catholic and Jeffrey's family being Baptist, which was an excuse for lots of differences between our households. But eventually I understood that my mother had different ideas on how to spend money than Jeffrey's mother did.

She gave me an example. "We think it's important to have nice shoes to wear to church on Sunday. Perhaps there would be some money left over for ice cream if you hadn't somehow lost one of your church-going shoes last week," she said.

After that, I put much more *priority* on storing my shoes safely in the closet.

I pull out this story more frequently these days, now that Herbie got a subscription to *Electronics Cooling* magazine. He clips ads for various cooling products and drops them on my chair, with a sticky note saying, "Let's use this!" One time it was a liquid cooling system, with hoses and pipes and heat exchangers and a tiny electric motor-driven pump. Another time it was a scholarly paper about the benefits of jet-impingement cooling. "You shoot high pressure air through a special nozzle at almost the speed of sound right onto the hot component!" he gushed.

Usually he forgets about these gizmos after I fix his thermal problem, either by moving the hot part to a cooler spot on the board or by adding an off-the-shelf heat sink. But the last time it wasn't so easy.

"We really, really want to use the fastest possible version of the microprocessor in Phase 3 of the AP project, the one that puts out 45 watts of heat," Herbie said.

"AP? Isn't that the new name for the human brain unit?" I asked.

"Yeah. It's now called the Anthro-Processor. We were getting flack from the animal rights people, who somehow got the idea we were using cat brains instead of human. Anyhow, if we use this processor, we can reuse 90% of the software that was written for another project that got canceled last year. There is an 8 watt RISC processor that does the same thing, but we'd have to write all the code from scratch."

"Forty-five watts!" I said, "On one chip? The old HBU processor board uses only 30 watts total. And that barely works with the wimpy fans we chose two years ago."

"Yeah, I know," Herbie said. "That vein in your forehead starts to pop out when I mention any component with more than 2 watts. So I did your homework by looking at some more magazines. And I found the answer. Since we have to retrofit the new board in the systems that are already out in the field, I thought we could just add a fan right where we need it, right on top of the processor. Check this out—a heat sink and fan all in one, that clamps right on top of a microprocessor!"

Herbie showed me this picture of a fan-sink (see Figure 2-15). The miniature fan/heat sink combination has been evolving since the early 1990s. As the processors for the personal computer (PC) have increased in power over time, small heat sinks were added to them, and then larger heat sinks. When that became impractical,

Figure 2-15 The fan-sink: giving today's most advanced microprocessors the reliability of a fan.

when, for instance, the sink became bigger than the hard drive, tiny fans were added to boost their performance. Some vendors designed heat sinks to match the shapes of off-the-shelf fans. Other vendors built custom-size fans to match the shapes of popular heat sink sizes. The most exotic suppliers tossed out the plastic housing of the fan body, and integrated the rotor into the heat sink itself.

The thing Herbie showed me was no elegant thing of beauty. It was a big hunk of aluminum coupled to a high-powered fan—lots of surface area and lots of air flow, a real brute-force solution to a hot chip.

Herbie saw the smile of appreciation spreading over my face. "The spec sheet for this one claims its thermal resistance is about 0.25°C/W. Our 45 watt processor would have a temperature rise above air of only about 12°C. If the incoming air is, let's say, 60°C, that makes the processor case temperature only 72°C. Its case temperature limit is 100°C, so that is plenty of margin. Maybe we could even run the fan slower and save power. And it fits in the space we have. What's not to like?"

The tear started to trickle down my cheek. "This is a wonderful advance in cooling technology," I said, my voice a little choked up. "It's just too bad we can't use it."

"Hey! Now what's wrong? You just don't like anything that *you* didn't pick!" Herbie said.

"That's true enough. But I have another problem with the fan sink. It's just not reliable enough to use in our system," I said.

"Not reliable enough? These are being used in high-end workstations by the thousands. And you told me that almost every PC nowadays has got some kind of fan-sink inside. If they break all the time, how could they get away with that?" Herbie said.

I answered, "They don't break all the time. But the fan does tend to break a lot more frequently than the microprocessor that you are clamping it to. If we were making a PC or a VCR, I'd day that it was OK."

"What difference would that make?" he asked.

"For one thing," I said, "when the fan seizes up in your desktop PC in the middle of downloading some songs from the Internet, and the processor starts to overheat, you can just turn the thing off. When your product is carrying 600,000 lines of telephone traffic, you are very, very reluctant to ever turn the system off. Customers get kind of unreasonable when that happens. So in the telecom business we put a much higher priority on things like reliability and redundancy than they do in the computer world."

Herbie tried to outflank me. "But if fans are so unreliable, how come we use them in the fan tray to blow air through the whole system? Huh? Huh?"

"Those fans aren't buried inside the box where they are hard to replace," I said. "First of all, we made sure the fans are redundant. There are more than enough, so that if one fan dies, there is still enough air flow to keep the electronics running. Then we added alarms to tell the customer that a fan was dead. Then we made it possible to plug and unplug the fans without turning anything off, so the customer could replace a dead fan without interrupting any service. We did all that not because we expected them to last forever, but because we expected them to break!" I said.

"Well, so what if the fan might fail?" Herbie said. "The other components on my board can fail, too, right? What's the difference?"

"The difference is big, like the difference between Mickey Mouse showing up at a kid's birthday party and a real rat showing up instead. The failure rate of a fan bearing is from 10 to 100 times higher than the failure rate of the processor chip. Odds are, the first thing to go on your board is going to be this fan-sink. And it will tend to fail at a much faster rate than any other board in your system."

"What do you mean by faster? Something is always the weak link in the system."

I looked at the spec sheet for the fan sink Herbie had brought. "I'm no expert in reliability theory," I said, "but that's never stopped me from spouting before. The reliability folks have a way of estimating the failure rate for an electronic assembly. They add up the failure rates of all the components that go into it. For example, your processor has a failure rate, based on historical data, of about 200 FITs (failures in time—with a time base of a billion hours). If you add up the FITs for everything on your board, you get about 3,800. The failure rate for your fan-sink is about 4,000. Just by adding that one part, you are more than doubling the failure rate for the whole board. If you do the math, you'll see that you have reduced the mean time between failures (MTBF), which is one crude way to define the average life span, from about 30 years to less than 15 years. That may be OK for a PC, which becomes obsolete after three years anyway. Is 15 years MTBF good enough for the AP system? I don't know. You have to see how it affects the reliability of the whole system. But if the reliability goal for your board has to be higher than 15 years, then you'd better not depend on adding this fan sink to it."

Herbie made a sour face. "Are you sure? This thing just looks so slick. Isn't there some way we could make it work?"

"I suppose if the fan were turned off most of the time, and you turned it on only when the temperature got too hot, maybe you could extend the expected lifetime of the fan. You could have a little thermostat that turns it on and off. But I think that a 45-watt processor would need to have the fan going all the time or it would quickly burn up. So that won't help," I offered.

"So how are you going to solve this for me?" Herbie said, slumping in his chair. "Going to tell me to move the processor to the bottom edge of the board again? That's not feasible this time."

"Nope," I said. "But I think that I can figure out an oversize heat sink that might just work for an 8-watt processor."

"Great. That only means redoing all that code."

"Sorry. I don't make up these reliability requirements. I always had the impression that they called it *soft*ware because it was supposed to be easier to change than hardware."

"Dream on, Thermal Boy," Herbie said. He went away muttering to himself, something about maybe he could make the thing more redundant by adding a second fan-sink and connecting them together with a thermo-electric cooler and a heat pipe.

By then it was lunchtime. I checked the cafeteria menu for ice cream. One has one's priorities, after all.

PUTTING THE RIGHT SPIN ON FAN COOLING

There is one drawback to writing in an entertaining style: sometimes it is hard for people to tell when I am joking and when I am being serious. Maybe that is because my jokes aren't all that funny. I prefer to think it is due to the fact that reality can be a lot sillier than the things that I make up just for laughs.

I wrote a technical paper and submitted the abstract to a technical conference selection committee. It was about how the direction of fan rotation could have an effect on component temperatures. I had observed that in some special circumstances component temperature depended strongly on whether the fans turned clockwise or counterclockwise.

I got a friendly e-mail back from the committee. "Thanks for the entertaining satire. Your submission really lampoons the important-sounding, yet largely trivial stuff that passes for technical research these days. Thanks for keeping us on our toes with your little jest," it concluded.

It took me a while, but eventually I convinced them my paper was not a joke. Fan swirl and rotation direction do matter, and when I presented my paper at the conference[1], I heard from several people who were glad I had explained this mysterious phenomenon. In my presentation I showed pictures of a real shelf in which the air flow in each card slot changed dramatically, depending on the direction the fans turned. Since component temperature depends on air flow, I could say that temperature depends on whether the fans turn clockwise or counterclockwise.

How can the rotation direction of the fan blades matter? I used to think the idea was just plain silly. Then one day I was confronted with a new shelf with the code name "2100." The air flow patterns in it just didn't make any sense. There were dead spots in it that were in exactly the opposite locations from where I had predicted them using CFD. I was forced to analyze the 2100 shelf in more detail than I had ever wanted to. It eventually came down to answering this question: What difference does it make which way the fan rotor is turning, when we all know that air comes out of a fan uniform and straight (see Figure 2-16)?

If you think about fans at all, and I think about them all the time, this is probably the image you have of an ideal fan. If you have a little experience with fans, and have a more realistic imagination, you also picture the dead spot in the center of the fan above the motor hub.

Instead of imagining an ideal fan, just remember the last time you were dumb enough to put your hand up close to one. The air doesn't come out nice and straight. It comes out in a swirling cone that looks more like the next picture (see Figure 2-17).

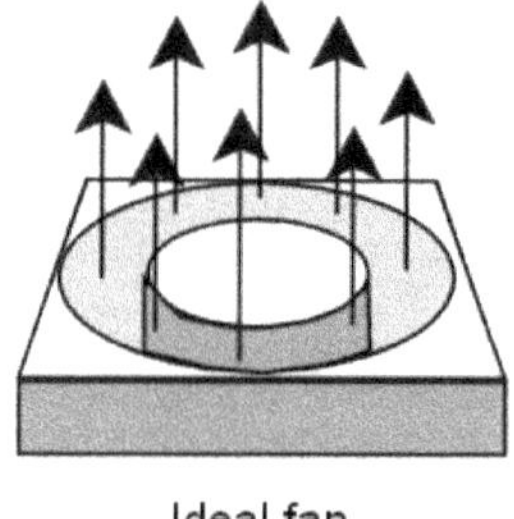

Ideal fan

Figure 2-16

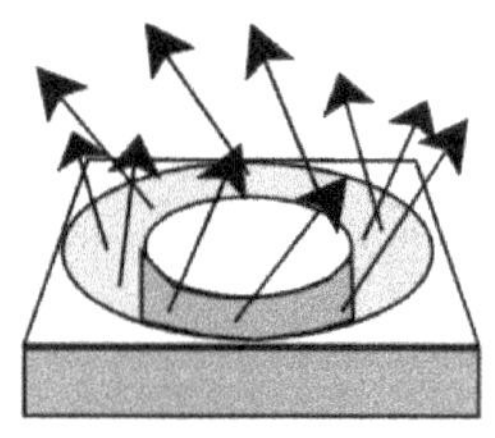

Swirly (real) fan

Figure 2-17

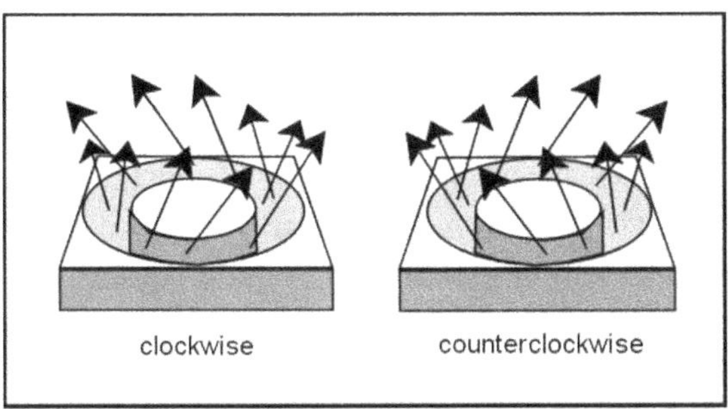

Figure 2-18

This is one of the times when air being invisible is bad. You can't see it swirling like my little arrows in the picture. But to satisfy yourself that turning, angled fan blades produce a turning, angled air flow, just dangle a small piece of twine near the outlet of a fan. You'll see the air does come out at an angle to the face, and that it is different all the way around. Not only that, but the angle depends on the direction of rotation. A clockwise fan has the mirror-image, opposite pattern of a counterclockwise fan. (I am not talking about taking the same fan and reversing its direction of spin. The air would flow backward through it, or maybe the fan would go backward in time. I mean a fan that is designed to rotate in the opposite direction, with blades facing the other way.) (see Figure 2-18)

Why hasn't anybody in the history of fan cooling noticed this before? I can't be the first.

Perhaps I was just lucky enough to encounter a shelf design that was so poor that it made the fan-direction problem obvious enough that even somebody like me would notice it.

This fan-direction problem is not very noticeable in typical shelves, because usually there are plenty of flow obstructions between the fans and the electronics, such as 90° turns or perforated plates. If, for example, your electronics are at the other end of a 10-foot-long duct from the fan, fan swirl doesn't matter. The flow resistance of the duct, grillwork or right-angle turn is enough to damp out the rotation effects. In cases like those, nobody *had* to notice the effect of fan rotation.

But the trend in electronic packaging these days is to crunch everything closer together, to make everything smaller, and to cram in more and more power. So fans are being added to boxes that never had them before, and the fans are getting closer and closer to the electronics. Sadly, there is no room in our racks anymore for the 10-foot-long duct that can cure many of the flaws in fan cooling.

I still haven't explained how fan swirl can really make a difference in electronic component temperatures. Instead of making fun of the poor innocent schmo who designed the 2100 shelf I talked about in my presentation, let's look at a hypothetical (i.e., oversimplified) example. It shares one critical design flaw with the 2100 shelf—the circuit boards are located very close to the exhaust side of the fan. What would happen if we placed an array of four circuit boards directly on top of the swirly fan shown earlier? The faceplates are shown cut away in this sketch, but in our little exercise, let's assume they make a nice, enclosed duct for air flow, with air flowing in the bottom and out the open top. The four boards form three air flow slots over the fan (see Figure 2-19).

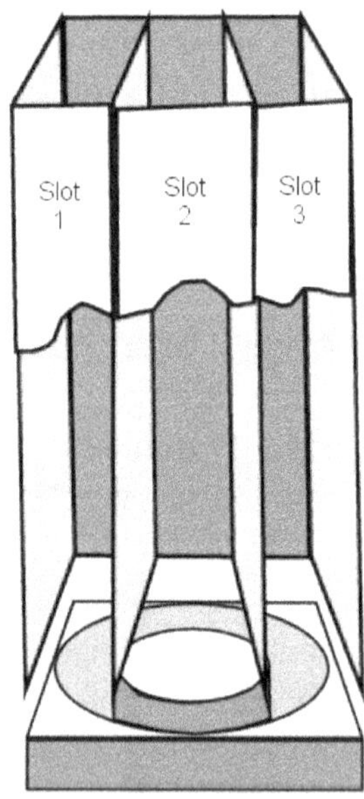

Figure 2-19 You'd get really good flow between these boards if you fit them right on top of a fan, wouldn't you?

Each slot is over a different part of the fan. What might the air flow pattern in each slot look like? Remember that the air comes out of the fan at an angle that changes as you go around the edge.

The next sketch makes it pretty obvious that the flow pattern is completely different in each slot (Figure 2-20). Slot 3, for example, has good air flow along the right edge, but a dead spot on the left. Suppose you figure this out by testing your little shelf. You might use this information when you lay out the components on the board used in Slot 3. You place all the hot components along the right edge, where you think they will get the best air flow. Good idea!

But then the fan tray designer, who might be way out in the woods of Minnesota and doesn't talk with you very often, finds a "better" fan in a new catalog. It has the same dimensions, better flow, less noise and costs less! He'll substitute it in a second without even realizing that it spins in the opposite direction of the fans you tested. (How would he know? Most fan spec sheets don't even tell you which way the rotor turns!)

With the new, "improved" fan, Slot 3 will now have good flow on the left and bad on the right. Your carefully designed board will get too hot. Unless you know that fan rotation even matters, you'll have a devil of a time figuring out why.

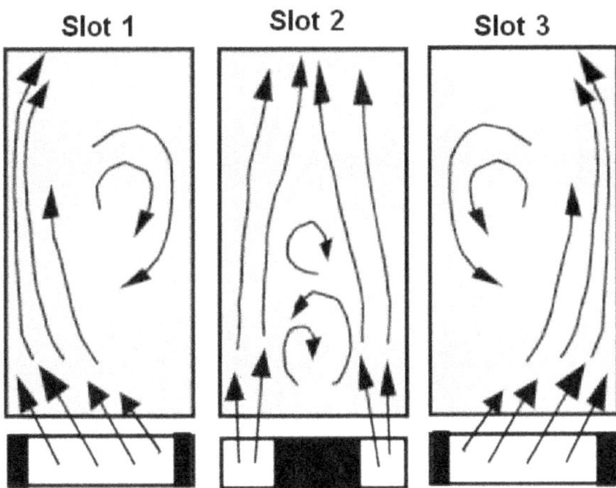

Figure 2-20 The flow pattern is different for each board, because the slot is over a different part of the fan.

That's why it's a good idea to put some distance *and* some kind of flow diffuser (like, my favorite, an air filter) between your boards and your fan. They smooth out the flow, more or less, and make fan swirl less troublesome.

Or you can put the fan on the exhaust side of the shelf (above the boards, blowing upward). There is no fan swirl on the inlet side of a fan. It's absolutely silly to think there might be swirl on the inlet side of a fan, so we should all just ignore it.

I think.

References

1. *"Fan Swirl and Planar Resistances Don't Mix," 9th International FLOTHERM User Conference.* www.flomerics.com

DEGREES C AND dBs

It sounded like Herbie on the phone. There was so much background noise that I could barely tell it was a human voice.

"Herb? Are you on an airport runway?" I shouted into the phone.

"What?" he said.

"Or are you chasing a tornado in a pickup truck?" I asked.

"Wasting a tomato?" he screamed. "Just meet me in the lab!"

Herbie was standing in the doorway when I made my way there. "I saved your hide again," he announced with a smug smile.

"Uh, thanks, I think," I said.

Herb continued, taking me by the elbow into the lab: "Remember how disappointed those marketing guys were when you told them the Seraphim shelf was not thermally feasible?"

I winced. They had looked disappointed enough to eat my stock options for lunch.

Seraphim was the codename for the optical interface for the HBU. Everything those days had to be optical, so we needed an interface between the HBU and optical fiber. And how else would a brain connect to beams of light, but with optic nerves and eyeballs? The first version, the ON&E (optic nerve and eyeball) module, with two eyes, was received so well that customers immediately demanded a higher capacity version. Somebody created a PowerPoint slide of an ON&E with eight eyes, so that proved it could be done, especially since it had already been shown to a customer. The many-eyed creature with the fiery power dissipation reminded somebody of an angel from the Book of Isaiah; hence the name Seraphim.

Seraphim was set to take flight, until I poked it hard, Moe-style, by pointing out that its fan tray did not put out enough air to keep it cool. The ON&E module barely worked with a few degrees to spare, and now the power was tripled. The thermal feasibility question was a no-brainer, in more ways than one. Without more air flow, Seraphim, and the HBU connected to it, would literally cook.

Herbie said, "You claimed the existing fan tray couldn't put out any more air flow unless it got bigger. And there's no room for that. But I found this fan on the Internet that's the exact same size as the one we already use, but twice as powerful! The company that sells it also deals in gold bullion and imported frozen catfish fillets, but this fan is a real breakthrough."

He handed me a fan out of a large cardboard box that smelled vaguely of fish (see Figure 2-21). The fan was wrapped in a data sheet. Herbie was right. Across the top of the "Omega Fan" sheet was the matter-of-fact statement, "This fan is a real breakthrough!" According to its performance curve, it would produce twice as much air flow for a given pressure than the fan we were already using.

Figure 2-21 The breakthrough Omega Fan, perhaps the last fan you'll ever hear.

"A data sheet is all well and good," I said, "but how do you know it can really…"

"Let me show you!" Herbie gushed, leading me toward his equipment racks. "I already popped six Omegas into an existing fan tray. It's running side-by-side with our original fan tray for comparison."

Herbie pointed to the two fan trays, mounted in racks. He gestured for me to put my hand over the exhaust vent of each. Gingerly, I did.

"Wow, that *is* a lot more flow than from the one on the right," I said. "A lot more noise, too!" The sound was similar to a blow-dryer running between my ears.

"Floor toys who?" Herbie shouted over the whirring.

I flipped off the breaker to the fan tray and the noise vanished. Gradually I became aware of people chattering at the next bench, phones ringing, fingers clacking on keyboards. "A lot more noise, too," I repeated softly. "There's a cost that comes with using a more powerful fan. To push more air, the fan has to spin faster. And the faster it spins, the more audible noise it makes. Didn't you know there is a Telcordia audible noise limit for equipment in telecom central offices?"

"As a matter of fact, I did," he answered. "GR-CORE-63 limits audible noise to 60 dBA. The Omega is safely under that limit at only 59 dBA. What's the problem?"

"It's not like the voltage rating. Noise sources add together!" I said.

Herbie looked skeptical, "You're saying two fans would be 118 dBA? That doesn't seem right. I thought that level was reserved for jet engines and boom boxes."

"It doesn't work quite that way. Since the loudest sound you can hear is about 10^{13} times more powerful than the quietest sound, it makes sense to measure loudness on a log scale. I could give you the official equation[1], but you'll only remember this shortcut anyway. For Sound Pressure Level, which is used in the Telcordia standard, every time you double the number of noise sources, you increase the loudness by 3 dB (dB is a decibel). So if one fan is 59 dB, then two fans would be about 62 dB. Four fans would be 65 dB, and your six fans together are someplace between 65 and 68 dB." I said.

"Big deal," Herbie said, "68 dB doesn't sound that high. I don't see blood coming out of anybody's ears. Just how high can you go before that happens, anyway?"

We found Table 2-1 on the Internet. It gives you an idea of where 60 dB is on the noisiness scale.

"According to the OSHA (the U.S. government Occupational Safety and Health Administration) Web site, you don't even have to start worrying about hearing damage unless the noise level is 85 dBA," Herbie said. "So why is Telcordia so persnickety about 60 dBA?"

"I'll give you a highly detailed explanation," I said. Then I turned the Omega fan tray back on. I continued my lecture through the torrent of noise. After a long minute, Herbie got the point, and shut off the fans.

"…normal conversation," I finished.

"OK, so fans at this noise level won't hurt your ears, but they make it pretty hard to carry on a conversation. And I suppose people have to be able to work together in the same room with our equipment. And if they can't understand each other, they might make mistakes and knock out phone traffic," Herbie reasoned.

"Maybe you could dampen the noise with baffles and foam, but that would take more space again. Or you could slow the fans down.

Table 2-1

Decibels (dB)	Typical Sounds
120	Artillery, rock concert
110	Elevated train
100	Boiler factory or TV commercial
90	Unmuffled truck
80	Noisy office (copy machine or your cube mate on the phone)
70	Average street noise
60	Average factory or home with kids
50	Average office
40	Library or funeral home

Here's how you calculate the noise change when the speed (RPM) changes:

$$\textbf{Noise change} = \textbf{50 log}_{10}\textbf{ (RPM}_2\textbf{/RPM}_1\textbf{)} \qquad \text{Eq. (2-7)}$$

If you cut the fan speed by 50%, you lower the noise of a fan 15 dB," I said.

"But if I reduce the speed, I'll just get the same air flow as your wimpy old fans," Herbie said, "The only way the Omega drives more air is by turning faster! Now what am I going to do? The marketing guy is on his way to see this!"

I thought for a second, then said, "You've got two choices. You could fake your own death. Or maybe you can convince the marketing guy that all this white noise is a new feature."

Notes

1. The official equation for adding together identical noise sources is:

$$\textbf{L}_N = \textbf{L}_1 + \textbf{10 log}_{10}\textbf{ (N)} \qquad \text{Eq. (2-8)}$$

Where N is the number of noise sources; L_1 is the audible noise rating of one noise source; and L_N is the total audible noise of all N sources. Check it out. If $N = 2$, then $10 \log_{10}(2) = 3$ dB, just like the rule of thumb.

WKUL-AM

I was doing my Sunday morning radio call-in show about Electronics Cooling ("***Hot Air On The Air* with Tony K**"), when I heard a familiar voice on the line.

"Herman in Homewood," I greeted the caller, "you're on the air with '***Hot Air***'."

"First time caller, long-time listener," the caller said. "I listen to you every week before I go to church…"

"I thought you were going to say the golf course," I interrupted. Johnny, the chain-smoking producer behind the glass, rolled his eyes at my attempt at humor.

Herman said, "Golf? People with electronic thermal problems tend to pray a lot more than other folks."

"You got me there," I said. "Now what's running too hot for you today?"

"I listen to your show, and I read your book and I'm planning to see your movie…" he started.

"Oh, baby, the movie! Do you think Lithgow can do me justice on the big screen? I wanted Hanks, but, in the audition, he just didn't seem all that passionate about heat sinks," I said.

"Lithgow will be much stronger in the dance numbers, no doubt," Herman said, "But back to your book—you constantly harp on one thing: component junction temperature. You say the whole point of cooling electronics is to get the silicon chip temperature down."

"Guilty," I said. "It's inside the components where the heat is generated that the temperature-related failures occur. You can't just measure exit air temperature and conclude your electronics are fine. You need to measure component temperature."

Herman continued, "OK, so I'm following your philosophy, concentrating on component temperatures, keeping them below their operating limits. But by keeping both of my eyes on junction temperatures, I've missed other thermal problems."

"Sounds like you have a particular example in mind," I said.

"I sure do," he said. "I'm working on an electronic pet-tracking system, combining cell phone and GPS technology. You implant an ID chip under your dog's skin, and if he runs away, the cops can home in on him. I'm doing the box that does the tracking and map display, and it needs the highest speed processor they've got. When it's tracking a really fast-moving dog, like a greyhound, that processor really hums, and kicks out 110 watts."

"That processor is no dog," I said. "So you want me to recommend a way of cooling it?"

"No, that's covered," he answered. "I clamped on a big heat sink, with a fan right on top. In this case it was simple to know the processor junction temperature, because it has a built-in temperature sensor, which you can read right on the computer screen" (see Figure 2-22).

Figure 2-22 Herman's heat sink-fan before the maximum ambient test.

"Sounds good. So what went wrong?" I asked.

"Everything looked good on paper," Herman said with a sigh, "Then we did the qualification test. If you've ever spent the night in the clink or dog pound (and who hasn't?), you know that they aren't exactly the Ritz when it comes to climate control. When it comes to animals and cops, the safety and environmental standards aren't very strict. They allow room temperature to be as high as 40°C (104F). In my bench test at normal room temperature of 20°C, the processor was 70°C, so I expected it to get up to 90°C during the maximum ambient test. That would have been OK, because its operating limit is 105°C. So I wasn't expecting any problems when we got around to testing it at its maximum operating temperature. We put it in an environmental chamber at 40°C. Everything started out fine. The processor temperature went up to 88°C fairly quickly, and then stayed there. Everything looked great. But after two hours, the processor temperature suddenly went way up very quickly and then the readings stopped. The processor had failed."

"That's weird," I said. "What happened? Was it a bad component? Did the heat sink fall off?"

"After we opened up the box, it was pretty obvious," Herman said. "The processor was 88°C, the heat sink was about 78°C and the fan was about 75°C. I found out, unfortunately, after the test, that the inexpensive plastic fan I had chosen was rated to only 60°C. The heat from the heat sink softened the plastic fan body, and when it warped, the fan stopped turning. Without cooling air, the processor heated up until it failed" (see Figure 2-23).

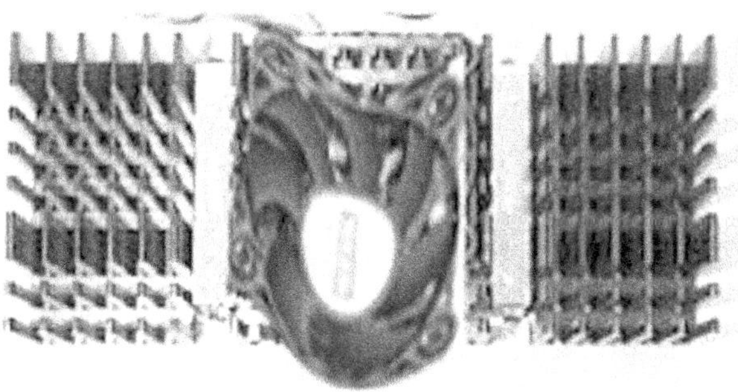

Figure 2-23 Turns out the fan had a lower temperature limit than the processor it was supposed to keep cool.

"Ouch!" I said. "Sounds like you need a fan with a higher temperature rating."

"Duh!" he reported, "I know that *now*. But I was busy worrying about junction temperatures, not fan temperatures. I didn't imagine a fan could overheat, until it was too late!"

The phone board was suddenly flooded (with three) more calls. Johnny started putting them on the line without asking me. He does that to bug me.

"Hi, this is Jane from Janesville. Herman is right. You can't just concentrate on the electronic component temperatures. Are you aware that for the European telecom industry, ETSI requires that exhaust air from a cabinet cannot exceed 70°C because it might blow into the cable plenum, where the electrical cable insulation is rated for only 70°C."

I tried to answer, "Yeah, but…"

"Tony K., this is Gail from Galena. My DSP chip was OK with a case temperature of 90°C, but we routed optical fiber from a laser transceiver over it, and the plastic jacket on the fiber melted. It was only good up to 85°C. And melting wasn't good, I found out."

"Gail, of course you need to…" I tried.

"Am I on? OK, this is Joe from Springfield. Don't forget about human contact temperature limits. That heat sink on the back of your power supply may keep your diodes happy at 100°C, but you sure don't want somebody to touch it. Temperature can be a safety hazard, so keep touchable surfaces less than about 70°. Your favorite telecom document, NEBS GR-CORE-63, requires that the aisle-facing surfaces of any equipment, like the front doors, have to be less than 38°C, when the room temperature is 26°C. So people walking by don't get an accidental tattoo of your company logo if they brush against it. Oh, and sorry, I couldn't think up a clever radio pseudonym. I really am just Joe from Springfield."

Herman, still on the line, added, "And what about the printed circuit board? Doesn't epoxy have some temperature limit? Maybe it's OK for a wire-wound power resistor to be 175°C, but not if it toasts the board. My typical board has a maximum use rating of only 105°C. You've steered us wrong, thinking only about junction temperature."

"Yeah!" the other callers chimed in.

"OK," I said, "I *may* have heavily emphasized junction temperature over the last few years. But I had to go overboard, at least a little, to compensate for the widespread myth that all you had to do was measure the exit air temperature and you were done."

"Don't whine, T-man, we still love you," said Gail.

"That's what this show is all about," I said, settling back into my radio voice, "love. Love and heat sinks."

Johnny cranked his finger in the air, signaling me to wrap up before the news.

"Thanks, Herb—I mean Herman—for your insightful call," I summed up. "We can always learn from our mistakes, but it's easier to learn from someone else's disasters, and your dog-finding machine meltdown is a classic. So here's the lesson we can think about during church—we have to know the junction temperature of our components. That is a must. But we can't forget about the rest of the things that the electronic system is made from: circuit board material, fans, electrical and optical cables, paper labels and even sheet metal. They are all components, too, and they have temperature limits, just like op-amps, diodes and memory modules. You might even think of the cooling air as one of your components. I wish I could give you a checklist of things to measure—but it wouldn't have everything you might need to worry about on it. For example, how hot is a chip allowed to get if you implant it in a dog's neck?

"So until next week, this is Tony K. saying, keep cool, and remember our station's slogan: 'There's no problem you have that can't be solved by a radio talk show host'. This show was sponsored by Grandma Bonnie's boron nitride particles. If you want to put a little zip in your thermal grease, don't forget Grandma Bonnie's."

This chapter first appeared in the August 2002 Issue of *Electronics Cooling*. It is reprinted with permission.

Section 3

Components and Materials: the Sum of the Parts is Sometimes Just a Big Hole

When university labs investigate heat transfer in electronic assemblies, they hardly ever use printed circuit boards and electronic components to represent printed circuit boards and electronic components. Boards are represented by uniform thermal insulators like balsa wood, and components are represented by blocks of aluminum. Why? Quite understandably, the researchers need to simplify the assemblies to make them easier to understand. They are looking for fluid flow and heat transfer phenomena, and they don't want a lot of the details of real electronic components to complicate the picture.

Too bad we can't do the same thing. It would be a lot easier to figure out component temperatures if they were all uniformly sized aluminum blocks. And conduction within the printed circuit board would be a lot easier to calculate if we didn't have all those messy copper traces mixed in with the dielectric.

But we who must cool electronics in the real world don't have that luxury. Electronic assemblies are not built with thermal needs in mind. So in addition to learning the arcane details of conduction, convection and radiation, you will have to learn something about electronic packaging and the properties of the unusual materials used in it. You will need to answer some questions you thought you'd never hear, such as, "What happens if you gold-plate an aluminum heat sink?"

NOT WORKING WITHIN THE LIMITS

Herbie was not happy with me, and he was enjoying it. "You told me that my DREC (dual redundancy encephalo-conduit) module would be OK thermally, but now it turns out I'm in deep trouble."

He showed me an old thermal simulation report. I skimmed through the component temperature predictions and shrugged my shoulders. "What's the problem?"

"These three custom chips, right here in the middle of everything, the Id, the Ego and the SuperEgo chips. You predicted the worst-case junction temperature would be 92°C, and you said that was OK because the component spec gave an operating junction limit of 100°C," he started.

"The last time I checked, 92 is still less than 100," I countered. "What's wrong, does it come in higher than that? Those things happen in this business."

"Nothing wrong with your prediction," he said, "just with your operating limit. That 100°C limit in the spec means that the chip just doesn't work at all. My trouble is that in my circuit the signal timing is really critical. And when the Id and Ego chips get over 80°C, they just stop talking to each other. They aren't damaged, and they still give out the proper signals, but the timing is off just enough that by the time one says something, the other one isn't listening anymore."

"What about SuperEgo?" I asked.

"That's only used in diagnostic mode," Herbie said, dismissing it with a wave of his hand. "It doesn't really do anything in the day-to-day operation of the system."

"80°C? Are you sure?" I said. "That is pretty low for an integrated circuit. How do you know it's temperature-related?"

"Everything works great when the system is all laid out on my bench at room temperature. But when I put the thing in our little environmental chamber, and start increasing the temperature, the circuit starts to act dumber and dumber. It's like it starts to get senile with higher ambient, instead of age," he said.

I shook my head and said, "This is one of the things I just don't get about electronic component specs. Whoever writes them sticks in a max operating temperature. A single number, like 100°C. But they don't tell you what it means. Does the die burn up at 101°C? Does it start to lose functionality? Or is it that the chip starts to become unreliable? There is no clue. So how am I supposed to know about this timing-problem limit? It depends on how good the timing has to be in your circuit. To a thermal guy like me, these chip look like just a bunch of oddly shaped resistors that spend all day changing electricity into heat."

This is where things can fall in the crack. Temperature affects the way circuits work, even within the so-called normal operating limits. But the circuit designer (Herbie) and the thermal engineer (me) each only understand half the problem. If we don't stick our brains together in just the right way so there is some overlap, these temperature effects may not be discovered until it is too late. He and I have to communicate at least as well as the Id and Ego chips if they are going to have a chance of working together.

Here are a few examples that I have been able to dig up of temperature messing up the function of a component, even when it is within its "normal operating range":

Capacitors

Chip capacitors are pretty good within the range given in their specs, but watch out for the cheesy Class III ceramic capacitors.

They can lose as much as 56% of their capacitance before they reach their operating limit of 85°C ambient.

Power Rectifiers

Figure 3-1 is the curve for the reverse current through one particular Schottky rectifier. When the junction temperature of the diode goes above 75°C, it starts behaving more like a resistor than a diode. The power dissipation of the reverse current skyrockets from 0.045 watts at 75°C to 4.5 watts at 150°C. That can be the recipe for thermal runaway. (Recipe for thermal runaway: Take one teaspoon of heat, and the temperature goes up. But if the higher temperature changes the characteristics of the device, so that it starts making even more heat, the heat and the temperature can quickly spiral upward, like some kind of crazy electronic soufflé.)

EPLD

In contrast to the previous part, this customizable electronically programmable logic device (EPLD) chip has the curious feature that

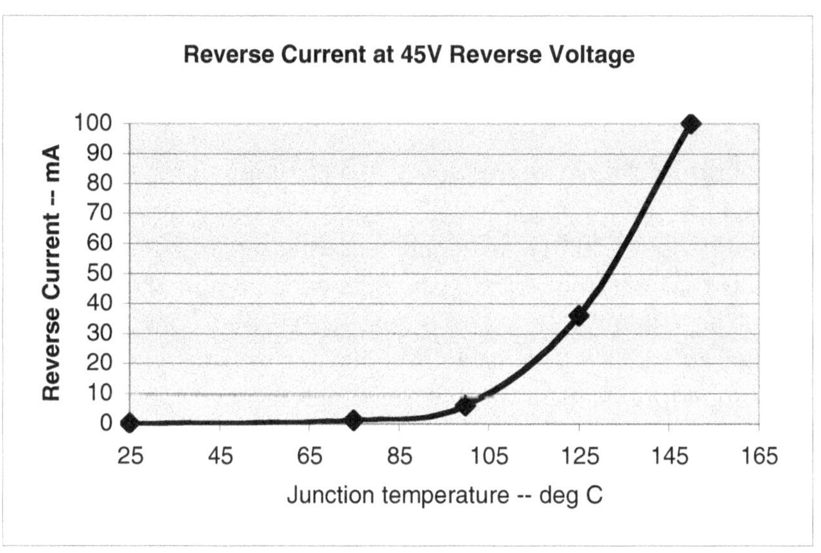

Figure 3-1

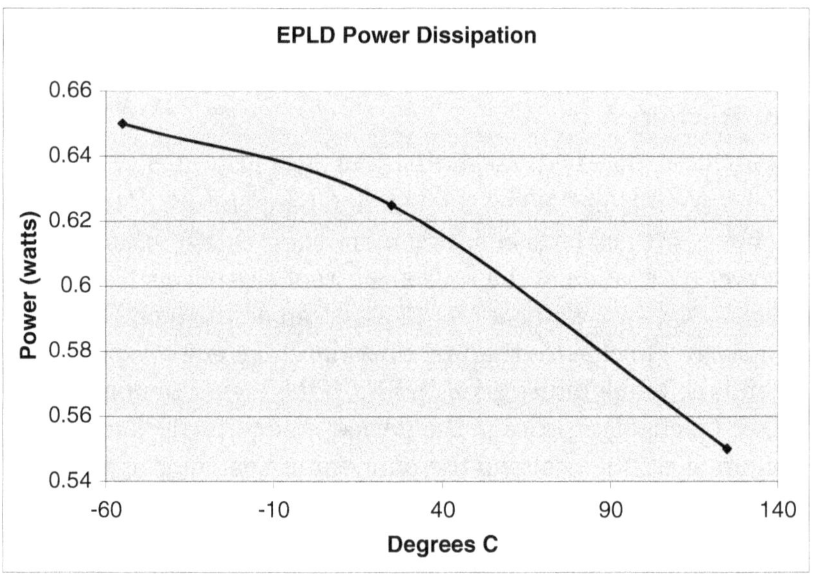

Figure 3-2

the hotter it gets, the less power it needs (see Figure 3-2). Herbie thinks that if he puts this part together with the diode above, maybe he could get the circuit to run with no power input at all.

Crystal Oscillators

Temperature shifts the frequency of a crystal oscillator (see Figure 3-3). Generic vendor specs usually say only that the frequency drift is limited to about ±0.01% over the operating range. What they don't tell you is that the frequency change depends on how they cut the crystal. These graphs show how a quartz crystal's frequency depends on temperature for different orientations and cutting angles. Obviously some types would work better at high temperature than others, if you care about the accuracy of your clock and timing circuits.

Batteries

Every winter, when I try to start my car in the morning I am reminded that batteries don't like the cold. They also don't like get-

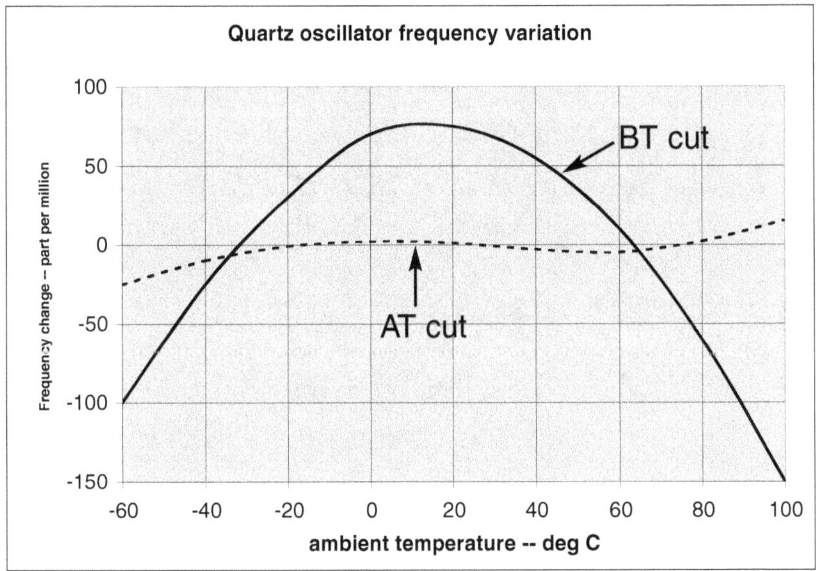

Figure 3-3

ting too hot. Almost all kinds of batteries have maximum energy density (that is, they hold the most "juice") between 30 and 40°C, a tad warmer than room temperature. Above and below that range, the energy you can get from a battery drops off as dramatically as your attention span in an after-lunch meeting. That applies to lead-acid batteries. Different chemistry batteries have different behaviors—but you can count on one thing: temperature affects all of them strongly.

DRAM Availability

Dynamic RAM is like a teenager—it needs to have its memory refreshed every few nanoseconds or it doesn't remember what you just told it. Availability is the time when you can actually read or write the memory, which you can't do while it is being refreshed. Refresh is necessary because current tends to leak away from MOS-based DRAM. That is not a problem at normal temperatures, when the availability is about 99%. But current leakage increases with temperature. Supposedly, when the junction temperature gets above

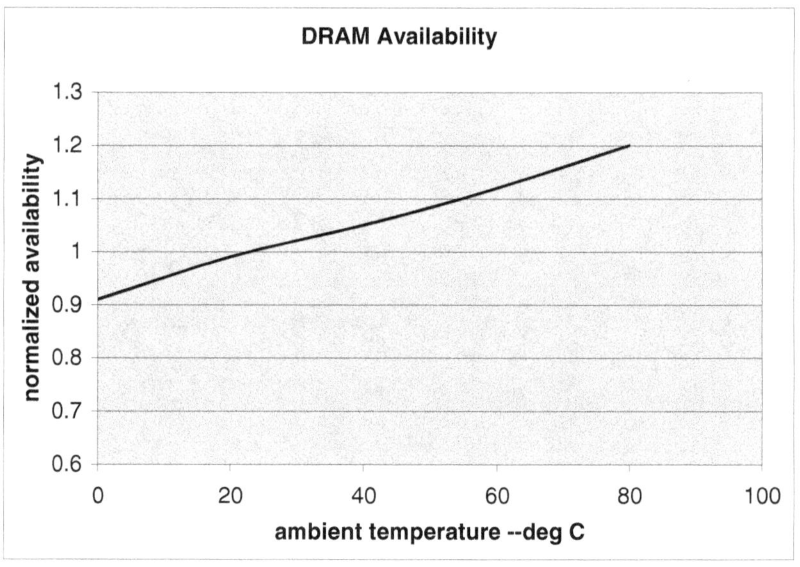

Figure 3-4

100°C, most of the DRAM's time is taken up just doing refresh. Data books are mostly silent on this topic (I wonder why), but I did find this graph (Figure 3-4) in a Hitachi book that shows access time increasing with ambient temperature.

It shows memory access time increasing about 20% over the nominal operating range. Military reliability handbooks say DRAMs can operate all the way up to 125°C junction temperature. Maybe they still have good reliability at that temperature, but what happens to the availability?

These are just a few examples of temperature effects I happened to find out about. They don't begin to cover the full range of components, and I am the wrong guy to be writing about this topic, anyway. I don't know how these things work on the inside. I just try to keep them cool enough. What about inductors, microprocessors or lasers? Do circuit board properties, such as the capacitance between layers, change with temperature? I have a feeling they do, but how important is that in the functionality of a typical circuit? What is the functionality of a "typical" circuit?

I had to give up on finding a funny, pithy, educational punch line to this story. The lesson learned is that Herbie and I opened up a big can of worms when we ran into this problem. It turns out that for many types of electronic components, there isn't just one, simple, operating temperature limit. Temperature and function can have a complex interaction, and that interaction changes when the component is applied in a different circuit.

The sad part is that the more complex the interaction of the device and temperature, the more complex the interaction between Herbie and me has to be. Which means my dream of moving to Hawaii and solving thermal problems via e-mail and instant messaging will be put on hold for the time being.

DON'T BLOW IT WHEN SIZING A FUSE

Almost every circuit board has at least one fuse (see Figure 3-5). The ones that don't probably have a component or skinny trace that acts like a fuse, although the designer may not have intended it.

The fuse was not invented by a circuit engineer. Engineers expect electricity to follow the schematic, and they know they wouldn't draw in any short circuits, so why would they bother even conceiving of a component to prevent a disaster in case one happens? Fuses were invented by fire insurance companies, for a very good reason. That is why selecting a fuse requires the engineer to think backwards.

Figure 3-5 The fuse: ironically, when you size it properly, it never performs its function.

Fuses are different from every other kind of component, because:

- If you select the proper size fuse for the application, it never does anything.
- It is the only type of electronic component that works strictly on thermal effect, yet we never bother to predict or measure its temperature.

Jacques called from the Montreal development lab with an update on the environmental testing on the OutreNet Project. He had designed its major circuit boards, the outdoor unrestricted interface (OUI), and the network optimization node (NON) modules. OutreNet was a multiterminal Internet access system to be mounted outdoors on city streets. It was a government agency project, a way of bringing the Internet to every citizen. "Home Pages for the Homeless" was its slogan.

The electronics were mounted inside a crowbar-proof kiosk, bolted to the sidewalk. The kiosk can be exposed to all kinds of weather, but to prevent tampering and infestation, there can't be any air vents. Its internal environment can get pretty toasty in the summer. The kiosk spec required Jacques's boards to work in a 60°C ambient.

For the high-speed, high-power components Jacques had planned to use, 60°C is a challenging environment. So I did plenty of thermal simulation of the boards for him before he built any. Guided by the computer temperature predictions, he shuffled components around, added heat sinks here and there, and substituted a few parts with higher operating temperature limits. Eventually, Jacques had layouts that we agreed should work inside the kiosk at 60°C.

"So how did your environmental testing go?" I asked, expecting the usual hearty congratulations.

"The test lab guys here are ready to strangle me!" Jacques said. "We have wasted half the day trying to get the system working long enough to finish the test. Every time we were about to complete the functional test cycle in the chamber at 60°C, the input fuse on the NON module blew. The first time, we cooled down the chamber, opened it up, replaced the fuse on the module, looked for some cause for a power short, like a bent connector pin, or a loose wire. We found a screw lying in the bottom of the card cage, so we assumed that was it. Then 20 minutes to boot up the system again, crank up the air temperature—and POOF! The same fuse pops again."

"Did you just put a penny in the fuse holder then?"

"Maybe I should have. But the second time, we replaced the NON with a new NON module. The system booted and ran fine, even after we got to 60°C. Then after 30 minutes of test signal traffic, the NON module fuse blew again."

I said, "Have you found your short yet?"

"This is a thermal problem!" Jacques said. "I finally decided to see if the fuse was the correct size. The measured input current was 2.5 amps at 5 volts. And the fuse is marked with a rating of 3 amps, 125 volts."

"Sounds OK to me. 2.5 is less than 3. Do you think your board draws more current when it gets hot?"

"That's not it. I measured the current at room temperature and at 60°C. It does not change. What changes is the temperature of the *fuse*. That fuse becomes so hot when the air is 60°, that it can't carry its rated current anymore. It melts."

"That's how a fuse works," I said. "It has electrical resistance, so as the current goes up, it gets hotter. When the current goes over the fuse rating, the metal inside must heat up enough to melt, and that breaks the circuit. It makes sense that if the fuse is used in a higher ambient, it would start out that much closer to its melting point."

Jacques said, "So when you did all those thermal simulations on the computer, why didn't you tell me that the fuse would get so over-whelmingly hot?"

That was a good question. I pulled out the NON simulation file. There was no fuse in the table of components. There was no fuse shown on the color temperature map of the board. I had ignored it. But the spot on the board where it belonged, smack-dab between two high-power Ethernet interface chips, was pretty hot—about 90°C. Any fuse in that spot would be at least 90°C, even if it wasn't gener-ating any heat.

I said, "It looks like we've got a *Catch Vingt-deuxiéme* here—you know, a Catch 22. When I do a temperature prediction on a board, I have to simplify. A board could have over 1,000 components on it, mostly tiny capacitors and resistors. The thermal simulator can only consider a few dozen components at most, so I toss out over 90% of them. Most don't dissipate any heat to worry about, so it is safe to ignore them. I include only components that produce significant heat: 100 milliwatts or more. Here's the trick with fuses: if you size them properly, they don't generate any heat. So it is safe to ignore them. But to size them properly, you need to know the fuse temperature. There's the catch. The only way I'd bother to find the fuse temperature is to assume you have chosen the wrong value. But for you to choose the right value, I have to tell you the fuse temperature. How bad of a fuse chooser should I assume you to be?"

After I did a little research, by reading up on fuse vendors' Web sites, I found out the situation was actually worse than I'd thought. Because a fuse works by melting a piece of metal, it is sensitive to the ambient temperature. The ambient temperature is not the air temperature of the room, but the local temperature around the fuse. The way I understand it, the ambient could be defined as the temperature of the fuse itself, if it isn't generating any heat itself. Thermal simulation could be used to find this "fuse ambient."

The nominal current rating of a fuse is only valid at an ambient of 25°C. A 5-amp fuse, for example, is rated to carry 5 amps only if the fuse ambient is 25°C. At lower temperature, it can carry more current, but the higher the temperature, the less current it can handle before blowing. This makes sense.

But it isn't even that good. Fuse manufacturers recommend derating the fuse to 75% of its rating, even at an ambient of 25°C, to avoid nuisance blowing. So even at normal room temperature, to handle a load of 7.5 amps, you should use a fuse with a 10-amp rating.

On top of that, you need to derate the fuse for its ambient. The following graph (Figure 3-6) gives an idea of how the current rating of a couple of typical types of "slow-blow" fuses changes with temperature. Don't use these specific numbers to choose a fuse in your application. The thing to take away from this graph is that different fuses

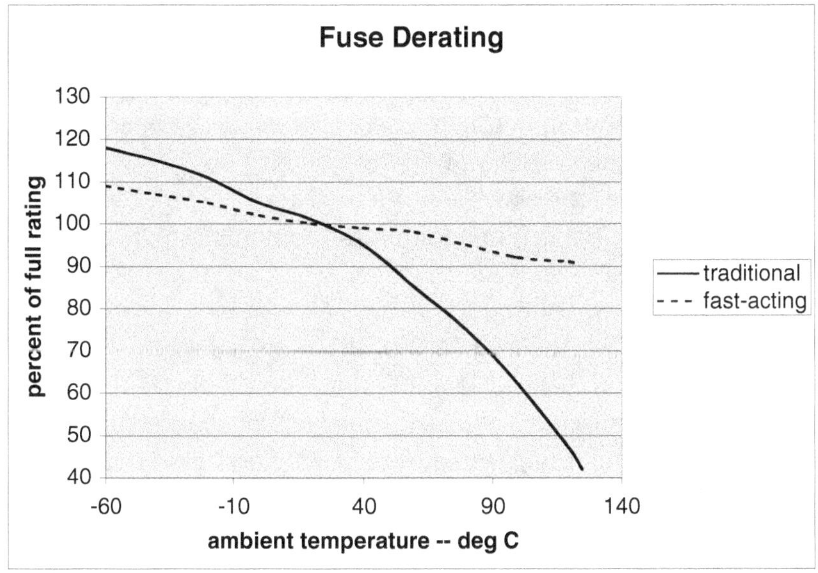

Figure 3-6

can vary with temperature quite differently. Some change a lot, some only a little.

Why are there two curves? The dotted line is for fast-acting fuses. Their current rating goes down only about 10% at an ambient of 125°C. The solid line is for traditional slow-blow fuses. Because they use lower melting-point metals, they are more strongly affected by ambient temperature. Guess which kind Jacques chose to use on the NON module.

The NON module draws 2.5 amps. At a nominal 25°C ambient, with a 75% derating, the fuse rating should be:

$$2.5\text{A load}/0.75 = 3.33\text{A fuse} \qquad \text{Eq. (3-1)}$$

You would probably round up to the nearest value of 3.5A in the fuse catalog. But Jacques had specified only a 3-amp fuse, which was probably too wimpy, even forgetting high temperature. Let's see what happens when we include the effect of a 90°C ambient on the slow-blow fuse that he chose; 90°C is the "local ambient" for the fuse on the NON module when the internal air temperature of the kiosk is at 60°C. He picked a slow-blow so it would survive the in-rush current of plugging the board into a "live" shelf, that is, so the board could be plugged in while the system power was still on.

At 90°C ambient, the slow-blow curve gives a current derating of about 68%. This is applied in addition to the nominal 75% derating factor like this:

$$\textbf{2.5A load/(0.75} \times \textbf{0.68) = 4.9A fuse} \qquad \text{Eq. (3-2)}$$

After rounding up, it looks like Jacques will need to change to a 5-amp fuse if he wants the NON to work at its worst-case temperature.

If he chose to use a fast-acting fuse (and deal with the in-rush current some other way), the derating for temperature at 90°C would only be 94%:

$$\textbf{2.5A load/(0.75} \times \textbf{0.94) = 3.5A fuse} \qquad \text{(3-3)}$$

Lessons learned from *Frére Jacques:*

- Derate fuses from nominal ratings (75%).
- Derate them again, based on the "fuse ambient" temperature. ("Fuse ambient" means the temperature of the printed circuit board where the fuse will be located. This may be a lot hotter than the ambient air temperature if nearby components generate lots of heat.)
- Don't automatically ignore fuses in thermal simulation, just because they don't dissipate power.

WHEN IT'S HOT, THEY ALL GO IN THE POOL

"Which is better," Herbie asked, "putting the heat from a component into the circuit board, or putting it into the air?"

I looked at him sideways as I lined up my next shot. We were out at the far end of the parking lot, shooting baskets during lunch, building up a protective coating of sweat for that afternoon's design review meeting. "What do you mean *better*? If you can get rid of heat by selling it to your grandma, do it."

Herbie pulled my missed shot out of the bushes. "I mean which works better? Seems to me that a circuit board, with all that copper in it, should make a pretty good heat sink. So if you could somehow connect up a component to the solid copper power and ground planes inside the board, you could get the heat out a lot better than by just letting it drift out into whatever air happens to float by."

"Sounds like you have something particular in mind," I said, watching him sink another 3-pointer.

"We're down-sizing the HBU, again. They're calling it the Head-Shrinker Project," he said.

"I always get a bad feeling when you tell me you're going to do the same function in a smaller space. Seems like most of the time you shrink the chassis by trying to toss out my fan system," I said warily.

"Not this time, my pessimistic buddy! We plan on shrinking things by cramming the functions of whole circuit boards into single custom chips. The only thing you might care about is the highest power chip will go from around 2 watts up into the range of 5 to 12 watts each."

"Whoa!" I said. My easy 6-foot shot went clear over the top of the backboard.

Herbie continued, raising his voice at me as I scrambled back up the embankment from the railroad tracks with the ball, "We want to pick a single package style for all these chips. We know we want a ball grid array (BGA) package, because we need lots of leads. But now there's a feud in the department between the "Cavity-Up" and the "Cavity-Down" camps. Since the argument is about which one is better for thermal performance, they need a voice from on high to settle it. But we'll also take your judgment" (see Figure 3-7).

I put the ball down and picked up a marker to draw on the whiteboard that some clever engineering manager had crudely clamped to the pole that held up the backboard. Sometimes when the weather was irresistible, we'd hold meetings under the hoop and we'd need something to write on. Other times a three-on-three game would devolve into a discussion of system architecture. It was hard to sepa-

Figure 3-7 A typical Cavity-Up style BGA with thermal balls under the die.

rate the schematic symbols from the basketball plays. I wiped them all away with a napkin from my lunch bag and started to sketch cross-sections of BGAs.

BGAs have become the rage in packaging. They increase the number of leads in the available space by putting the connections *under-neath* the body, instead of just around the perimeter of the package, as in your standard surface mount ICs. The leads aren't exactly leads either, but balls of solder, arrayed in a grid pattern, that connect the circuits in the package to the traces on the printed circuit board. That's where they get the name ball grid array [although a grid and an array are the same thing, so it is a redundant name, like Table Mesa or Gobi Desert (Gobi means desert in Mongolian)].

There are a variety of constructions for BGAs, but from a thermal standpoint, there are two basic types: Cavity-Up and Cavity-Down.

A Cavity-Up BGA has the die on the top side of its substrate (see Figure 3-8). The substrate is like a small, multilayer printed circuit board itself. For heat to get out of the die, it has to pass through the epoxy encapsulant to the air, or through the substrate, into the solder balls and into the board. Neither path is particularly good for conducting heat. Even when the heat can get to the solder balls, they mostly connect to incredibly thin copper signal traces, so the thermal performance of this package is not especially good.

A "thermally enhanced" version of this package adds solder balls directly underneath the die (see Figure 3-9).

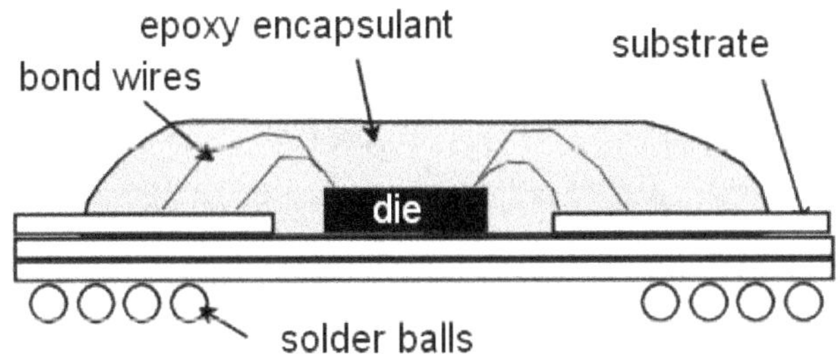

Figure 3-8

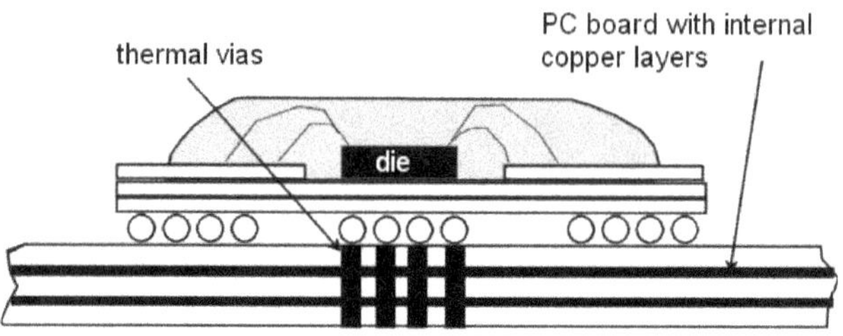

Figure 3-9

If the PC (printed circuit) board designer includes vias for these thermal balls, there will be a somewhat better path for heat from the die into the copper planes.

A Cavity-Down BGA may look similar on the outside, but inside it is upside down (see Figure 3-10). The cavity, (the hole in the substrate where the die sits), faces down, or toward the circuit board. In this version, the back of the die is adhesively bonded directly to a large copper heat spreader.

Heat from the die easily spreads into the copper plate on the top of the package, but the path from the die through the substrate and into the board is very poor. The cavity-down BGA relies on heat escaping into the air, rather than spreading into the board.

"OK," Herbie said, "so which one works better?"

"I don't know," I said. "I think it depends on where you're going to use it. Let's look at the Cavity-Up. It is optimized for spreading heat into the PC board. That works great if you have only one or two of them on a board. But what if you have 10 of them on a board, all bunched together?"

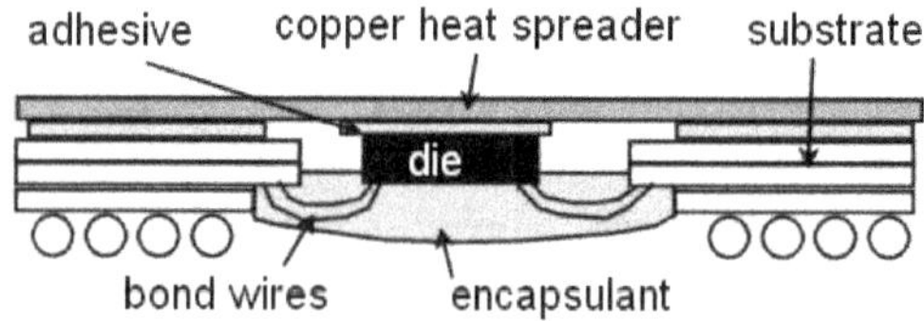

Figure 3-10

"I don't get it."

"It's like peeing in the pool. If you have one little kid letting loose in an Olympic-size pool, it spreads out pretty quickly and nobody notices. But if that pool is jammed shoulder-to-shoulder, like the old Humboldt Park pool I used to go to when I was a kid, and everybody is peeing at the same time (which they probably did back then), then there's going to be a noticeably uncomfortable temperature and color change."

Herbie squirmed at the image. "So Cavity-Up should be great if I don't put too many of them on the same board?"

"It all depends on how hot those copper planes in the board get. Suppose you have only one Cavity-Up BGA on a board, but it is right next to a Power Supply Brick. If the brick heats up the power planes, that heat will come right up through those thermal vias and balls and heat up your die," I said.

"So Cavity-Down is better, especially if the board is crowded?"

"Not necessarily," I said. "The Cavity-Down BGA counts on being able to transfer all its heat to the air. But what if the nearby air is stagnant, because you have a bad fan tray design? Or maybe the air is already hot because it passed over a bunch of other hot parts before it got there. The Cavity-Down kid is just peeing into a different pool than the Cavity-Up kid. The only thing I can tell for sure is that if you need to add a heat sink, it will be a lot easier to do it on the Cavity-Down package, because of its flat top."

Herbie handed the basketball to me for my next shot attempt. "But the data sheets say that the thermal resistance of the Cavity-Down BGA, you know, θ_{ja}, is lower than for the Cavity-Up BGA by a couple of degrees C per watt. Isn't that significant?"

My shot clanged off the side of the rim and nearly hit a parked car. "You already know how bogus θ_{ja} is (see "**Section 5. Tales of the JEDEC Knight**"). Comparing the thermal resistance of two packages is like comparing the fuel economy of cars. Your four-wheel-drive pickup gets fewer miles per gallon than my Toyota Tercel, right? So you'd say, based on that, that my car is more efficient at providing transportation. But what about on a muddy road? How many miles per gallon would I get in my car with the wheels just

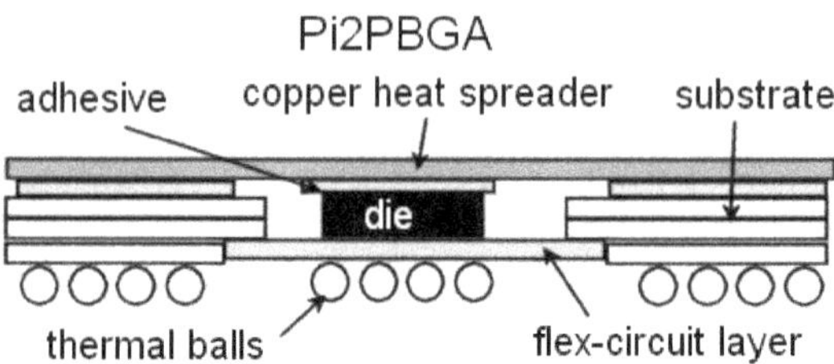

Figure 3-11

spinning in the same mud puddle all day, while your truck slooshes by and splatters me? Each vehicle is optimized for a different road environment."

Herbie spun the basketball on the end of his finger. "So how do we decide what package to use? It sounds like neither one is clearly the best."

I shrugged and offered, "I think it's going to take detailed computer modeling of the whole board layout to answer that question. You've got heat from multiple sources going in multiple directions. How can I give you a simple answer about which package works best in every possible situation? Maybe Cavity-Up will work great on some board designs, and maybe Cavity-Down will work better on others. Maybe neither one of them will work if the total power dissipation for the board is too high."

"Sounds like you don't want to pick sides in this feud. You can't be on both sides at the same time," he said.

"Both sides?" I said, "That gives me an idea."

I drew this up on the whiteboard (Figure 3-11).

"Tell your feuding camps that this is the kind of BGA endorsed by the thermal expert, if they can find such a thing. It has a low-resistance heat path into a copper spreader on top, *and* thermal balls under the die to conduct heat into the copper layers in the PC board," I said.

Herbie squinted at the card in the strong sunlight. "What is Pi2PBGA?"

"It stands for the **Pees-in-2-Pools BGA.**"

BYPASS CAPACITORS?

There is a reason behind the apparently mindless process that forces employees to move from one office to another every 18 months or so. Every time I have to move, I throw away half my accumulated junk. The same with work-related habits. I have to relearn where the nearest washroom is, which coffee machine dispenses anything drinkable and whom to bribe to book that nice conference room.

The company moves people at random on purpose. It costs hundreds of thousands of dollars a year, but it's worth it just to break us out of our ruts. Sometimes it's good to be forced out of your comfortable way of doing things, because it makes you think about your job instead of just turning the crank one more time. When you think, you might just come up with a better way of doing it. It's a lot more effective than forcing all the employees to take process improvement training.

That kind of accidental improvement happened recently when I tried out a new method of doing thermal simulation. The old crank-turning procedure was for the hardware design engineer to print a life-size drawing of his circuit board layout, and then I would hold a ruler up to it, get the location for every single component and type it into the Therminator software. It was tedious and slow, but it was reliable (until the little numbers began to rub off my ruler).

Instead of buying me a new ruler, my boss approved upgrading to Therminator Gold, which includes the mechanical computer-aided design (MCAD) interface package. This software, through the clever manipulation of acronyms (like IGES and ACIS), grabs geometry data from board layout or mechanical design computer-aided design (CAD) systems, and stuffs it directly into Therminator format. This not only saves me the time and eyestrain of creating the geometry over again, but prevents copying mistakes (although it does perfectly *preserve* all the original mistakes put in by the printed circuit board designer).

Herbie watched over my shoulder as I tried out the MCAD interface on one of his boards. He read me the instructions step-by-step and I mouse-clicked on the various icons on the computer screen. "Step 41-d," he read, "Select 'import'. Get coffee. Upon return, file will be imported."

Herbie's board eventually appeared on the screen, in all its glorious 3-D detail, components and all (see Figure 3-12). "It works!" I exclaimed in half-surprise. "Now instead of spending an hour typing,

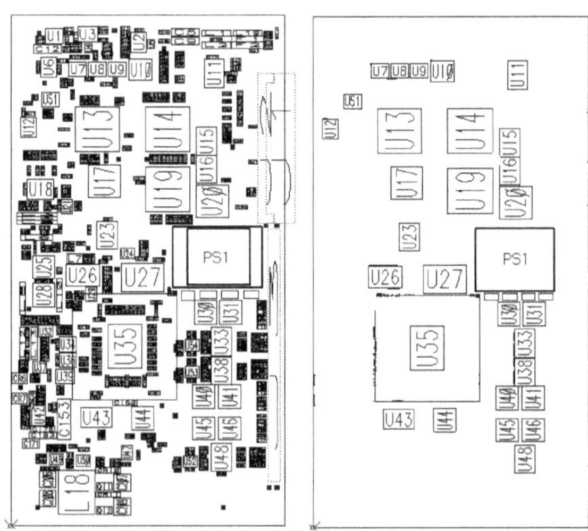

Figure 3-12 Herbie's board, before and after erasing "non-thermal" parts

I'll spend only 45 minutes erasing the 90% of the geometry that I don't use in thermal analysis. That's 15 minutes saved!"

"Erase?" Herbie said. "I spent weeks putting all that stuff where it is. Why do you need to erase anything?"

"The more details in the geometry, the longer it takes Therminator to analyze. So I get rid of any geometry that isn't important to the thermal behavior. Such as these mounting holes in the corners, these breakaway tabs and the 697 tiny pull-up resistors on the back side," I explained.

"Really? Don't resistors make heat?" Herbie said.

I shrugged as I clicked away with the mouse, making resistors vanish like barbecue chicken at a church picnic. "Usually it's safe to ignore them. They don't carry much current, and if you size them right, they don't get very hot. And resistors can withstand temperatures higher than most other kinds of components on the board. Besides, there are too many of them. Do you want to calculate the power dissipation of all 697 of them so I can enter it into the Therminator?"

"Nope," Herbie said.

Once the pint-size resistors had disappeared, I started in on the itsy-bitsy capacitors scattered over the board.

"Whoa! Hold on there, Thermo-Boy!" Herbie burst out. "Now you've gone too far!"

"What?" I protested. "They're just bypass caps. Same story as the pull-up resistors. There's a gazillion of them on the board, they don't dissipate any heat, and they're not sensitive to temperature."

"RRRRRRRN!" Herbie buzzed like the wrong-answer buzzer on the game show *Name that Brand Name!* "Wrong! Not only can a capacitor be temperature-sensitive, but capacitors *do* dissipate heat."

I was in shock. Too many new things in one day—a new software package to learn, a new procedure for creating geometry for the Therminator—and now Herbie telling me I was wrong about something instead of the other way around.

I said, "I always thought capacitors just charge and discharge. Energy goes in, gets stored for an instant, then comes out. So nothing gets converted to heat. Not like a resistor, where the electrical energy is all converted to heat. So you could always ignore capacitors in thermal analysis, since they don't generate any heat. And except for electrolytics, caps have pretty high operating temperature limits. When I created Therminator models the old way, with the hard copy and the ruler, I just automatically left them out without thinking about it."

Herb smiled and shook his head. "See these caps next to the power converter brick? Those are tantalum capacitors. I definitely want to know how hot those babies are going to get."

"Why?" I asked.

"They filter the output of the 5-volt power brick. That gives me a nice, clean 5 volts to power all the high-frequency stuff, like the processor, the DSPs, the FRZZ-TAG modulators and whatnot. And already there isn't enough room to fit everything I need on the board, so I want to use the smallest possible caps that will still work. The power brick will likely get pretty hot, and since they are right next door, the caps will get hot, too."

Then Herbie pulled out a capacitor catalog and showed me this graph of Working Voltage vs. Temperature (see Figure 3-13).

"The spec says tantalums are good up to 125°C. But look what happens to the working voltage when a tantalum gets over 85°C. It starts going down pretty fast. So fast that by the time the cap that is supposed to handle 6 volts gets to 125°C, it can really only handle 4

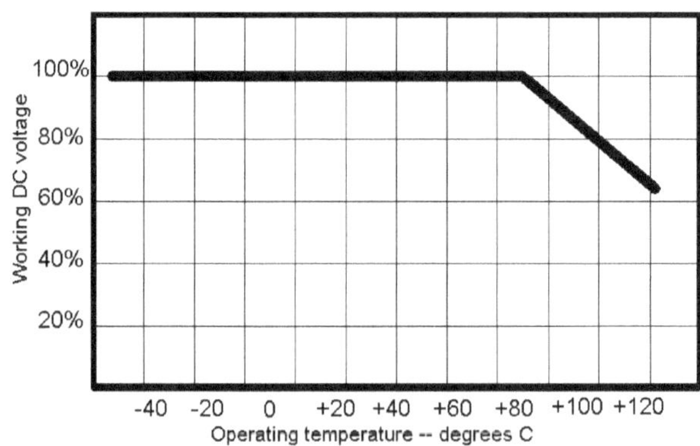

Figure 3-13

volts. So if I use that 6-volt cap to filter my 5-volt brick, at 125°C it'll fail."

"So just use a 10-volt cap! Or go hog-wild and stick in some of these 16 volters," I suggested, paging through the unusually-well-documented specs.

"That would work," Herbie said, "but the higher-voltage caps are physically bigger. And I already don't have enough room, so I don't want to waste space on extra-big caps if I don't have to. So this time, could you figure out the temperature of these capacitors and not just ignore them per your standard operating procedure?"

"As you wish," I said. "What about all these others, like these tiny capacitors on every lead of the microprocessor."

"Blow 'em away," Herbie said.

"OK," I said, "now I need to assign power to each component left on the board. These tantalums are zero power, right?"

Herbie pointed out another picture in the catalog (see Figure 3-14).

"Here's a schematic of how a real capacitor behaves. They can't help having some resistance and inductance built in. There is no energy loss in the capacitive and inductive parts, but some energy does change into heat here in the lead resistance and the dielectric loss. The heat generation depends on the frequency and the ripple voltage, so I'll have to estimate that for you. I know when you see anything with an Hz in it, your eyes glaze over."

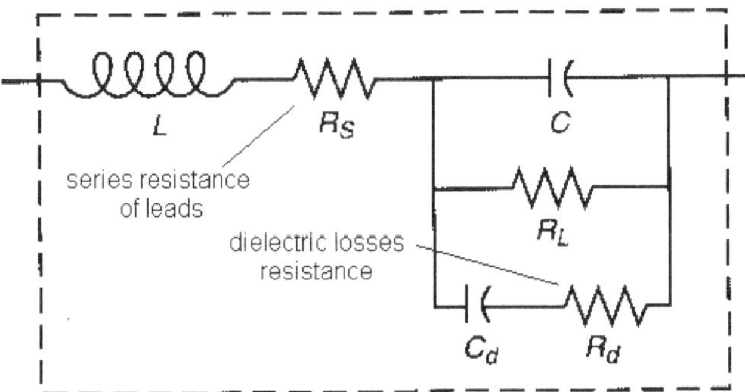

Figure 3-14 When a capacitor is more like a resistor

"Just give me the watts, Mr. Ripple Voltage, and I'll give you the temperature," I said, a little humbled.

"OK," Herbie said, walking away with a smug look on his face. "And watch those silly assumptions about capacitors from now on! It's a good thing I was here watching you build that thermal model so I could catch you in your over-simplification."

It stung a little, but eventually I appreciated that it was a positive thing to have to learn a new way of doing the same old thing. The ride may be bumpier when you are forced out of your rut, but you get the chance to find a brand-new destination.

A BAFFLING TEMPERATURE RISE

"What is the melting point of polycarbonate?" Herbie asked. As usual, it was the wrong question.

I wanted to say that plastics don't have a well-defined melting point, they just get less and less stiff as they get hotter, until they start drooping, long before they become liquid. But Herbie said, "For once, can you just give me the simple answer, not starting with 'when dinosaurs ruled the earth?'"

"But you love dinosaurs!" I said. "OK, OK, tell me the whole story. What are you doing with polycarbonate that it might melt?"

He sighed and told me, "It's another variation on the HBU project again. Our research scientists saw in *Reader's Digest* that people use only 10% of their brains, and were inspired to increase the bandwidth of the HBU. By eliminating random thoughts about stock prices and reproduction in the HBU, they were able to free enough capacity to carry the thoughts of up to 10 other brains, using a new protocol called brain wave division multiplexing (BWDM). That created demand for 10 times the number of telepathic port modules (TPM). So now it's my job to pack as many of these circuit boards into a rack as possible.

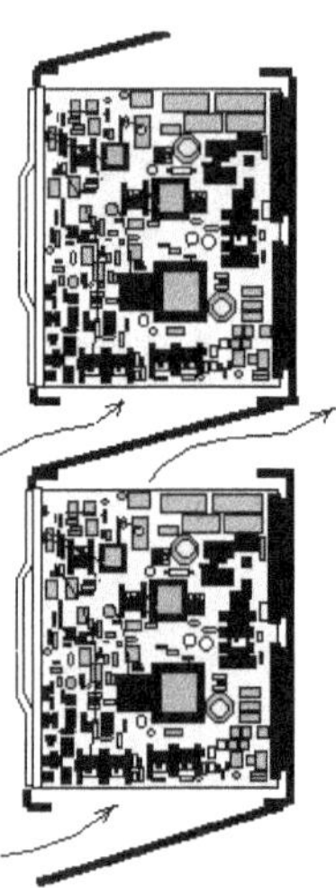

Figure 3-15 A baffle deflects the hot exit air from the lower shelf so cool air can enter the upper shelf. The smaller the baffle opening, the more air flow is choked off, and the hotter everything gets.

"I had to ignore your advice and make the baffle height between shelves as small as possible. It's only 1.75 inches high. That allowed me to cram eight TPM shelves into the rack, instead of the five or six I get using your baffle design. It barely works if you have only one or two shelves in the rack. But there's a problem with eight fully-loaded shelves" (see Figure 3-15).

Herbie continued, "Normally the baffle is a good barrier to heat, so each shelf is sucking in fresh, cold air from the front aisle, no matter how many shelves we stack up. But my very restrictive baffle chokes off the air flow so much that the exit air temperature for a single shelf is about 35°C hotter than the inlet air. So the bottom of each baffle is about 35°C hotter than the top. Some of that heat conducts through the sheet metal, pre-heating the incoming air for the next shelf. I measured the incoming air to the second shelf. It was about 2 to 3° hotter than the room air" (see Figure 3-16).

I shook my head side-to-side, disapprovingly, like Jessica Fletcher listening to a murder confession, and said, "I guess that would happen with such hot exit air. But 3°C doesn't sound that bad."

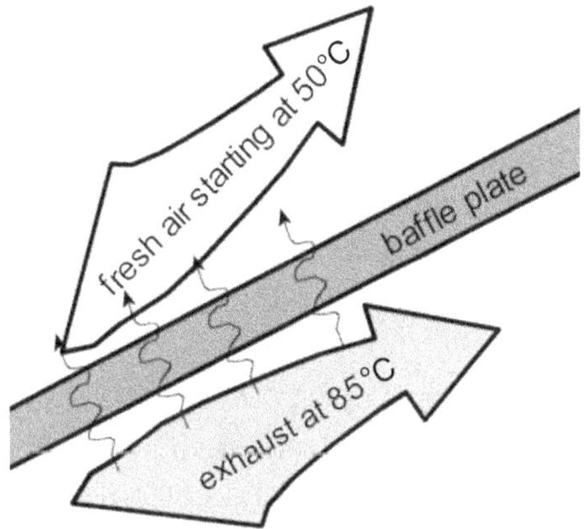

Figure 3-16

"Use your brain!" Herbie said, exasperated. "The problem is that the heat conduction through the baffles *builds up* as you go up the rack. The top shelf in the rack ends up being about 6°C hotter than the bottom shelf. And that's enough to put some components, like the neural nets made out of real neurons, over their operating limits."

"Talk about your fried nerves," I said. "So, how does polycarbonate fit into this picture?"

"The standard baffles are made out of sheet steel, which, being of the metallic family, is a pretty good conductor of heat. If I make them out of an insulating material, like plastic, I might stop the heat leaking through the baffles by conduction, and the top shelf would be just as cool as the bottom shelf. According to Chapter 4 of *Hot Air Rises and Heat Sinks,* the thermal conductivity of plastic is about 250 times lower than steel. So if I make the baffle out of polycarbonate instead of steel, the temperature rise across the thickness should only be 1/250th of what it is now, which would be practically nothing."

I said, "Sounds logical. But the polycarbonate might melt, a reasonable engineering concern. Tell you what—I'll look up the melting point of polycarbonate. Meantime, you can borrow this sheet of polycarbonate I've been saving to build my Natural Convection Wind Tunnel. Make some baffles on the band saw and try them out."

While Herbie sawed and duct-taped and measured temperatures, I did some thinking. Could Herbie have actually thought up a good idea? I scratched a few calculations on the back of an old quality systems training certificate and waited for his return.

"There's something wrong with your polycarbonate!" Herbie said later, brushing plastic crumbs from his flannel shirt.

"Let me guess," I said, "the plastic baffles didn't make a lick of difference."

"How did you know?" he gasped. "With the steel baffles the top shelf was 6°C hotter than the bottom shelf. With the plastic baffles, the top shelf was 5.9°C hotter than the bottom shelf. That's about 5.9°C less improvement than I was hoping for!"

"I checked your math," I said. "While I was looking up the properties of polycarbonate, it occurred to me that you forgot a couple of other thermal resistances that might affect how the baffle conducts heat."

"Like what?"

I showed Herbie this sketch for a book I am thinking about writing, *Boundary Layer Theory for Dummies* (see Figure 3-17). Believe it or not, when air flows over a surface, a very thin layer of air sticks to

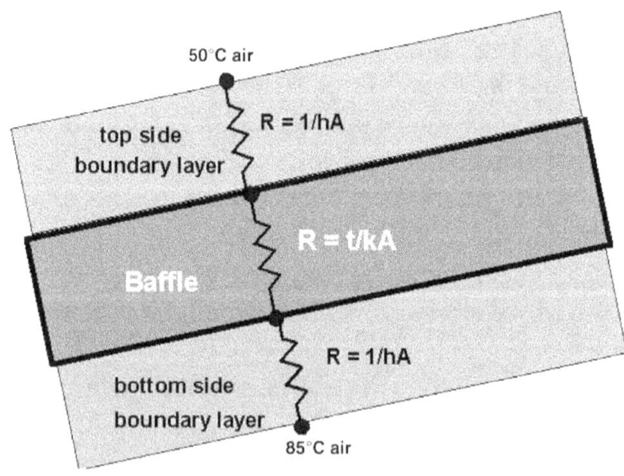

Figure 3-17

the surface like a film, and acts as a layer of insulation. In heat transfer books it's called the boundary layer—but you can think of it as *air slime.* For heat to transfer from the surface to the air, it has to pass through this sticky layer of air first. For heat to go from the hot air under the baffle to the cold air above the baffle, it has to go through three resistances in its path, like this:

The resistance of the air slime is:

$$R_{slime} = 1/(h \, A) \hspace{3cm} \text{Eq. (3-4)}$$

A is the surface area of the baffle and *h* is the heat transfer coefficient, which depends on the speed of the air flowing by. For natural convection over a flat plate like our baffle, the value of *h* is somewhere around 5 $W/m^2{}^\circ C$. The area of the baffle is 0.1 square meters. That gives a slime resistance on either the top or bottom side of the baffle of

$$R_{slime} = 2.0^\circ C/W \hspace{3cm} \text{Eq. (3-5)}$$

What is the resistance of the material the baffle is made from? For conduction through a solid

$$R_{baffle} = t/(k \, A) \hspace{3cm} \text{Eq. (3-6)}$$

t is the baffle thickness and *k* is the material conductivity. Table 3-1 compares the values for the two baffles Herbie tested:

Table 3-1

	Steel Baffle	Polycarbonate Baffle
Thickness *(t)*	0.062 in (0.0016m)	0.125 in (0.0032m)
Conductivity *(k)*	50 W/m°C	0.20 W/m°C
Area *(A)*	0.1 m^2	0.1 m^2
R_{baffle}	0.00032°C/W	0.16°C/W

Compared to the two slime resistances, even the conduction resistance through the thickness of the polycarbonate is very tiny, and by tiny, I mean so tiny it should be forgotten. The total resistance for the steel baffle is

$$\mathbf{R}_{\text{total, steel}} = 2.0 + 0.00032 + 2.0 = 4.0°\text{C/W} \qquad \text{Eq. (3-7)}$$

The total resistance after you change the baffle to polycarbonate is

$$\mathbf{R}_{\text{total, plastic}} = 2.0 + 0.16 + 2.0 = 4.2°\text{C/W} \qquad \text{Eq. (3-8)}$$

which is only 5% higher, even though you changed the baffle conduction resistance by 50,000%.

"I might as well stick with my old steel baffle. Making it out of an insulator doesn't seem to help," Herbie concluded.

"It would if you made it thick enough. For the plastic baffle to have the same resistance as one layer of air slime, you'd have to make it about 1.5 inches thick. But I don't recommend it," I said.

The same principle applies to the walls of any electronic box. If the air flow is fairly slow, for example, if you don't have fans, then the boundary layer (air slime) resistance dominates the heat loss through the walls. So if you have thin walls (less than 1 inch), it doesn't matter what you make them out of—plastic, steel, aluminum—or whether you paint or anodize them. The temperatures inside will be the same.

Herbie said, "If you knew that, why did you have me make all those polycarbonate baffles and do a useless test?"

"You always believe me more with a little test data," I said. "Plus, by coincidence, the panels you cut for your baffles are just the size I need to build my wind tunnel."

24K GOLD HEAT SINKS: WORTH THEIR WEIGHT IN ALUMINUM

Hyperactive was the first word that came to my mind when Herbie introduced his friend Roy. During our meeting he kept popping from the chair to the edge of the table, to the file cabinet, sniffing through my books and papers like a beagle.

Herbie tells "Roy stories" the way I tell "Herbie stories". Like the time Roy rigged up a garage-door-opener motor to his living room drapes so he could operate them by remote control. It worked great, if you didn't mind the drapes opening unexpectedly when the neighbor got home from work.

"Tell him about your invention," Herbie prompted, "A clothesline that doesn't need clothespins!"

Roy said, "It's barbed wire. Just toss the clothes on it and they stay by themselves. And the neighborhood kids never mess with it. Anyway, I need your advice on a heat sink."

"What hardware project are you working on?" I asked.

Roy smirked, "Oh, I don't do hardware here at work. My official job is Software Architecture Methodology Process Improvement Statistical Documentation. But one of my hobbies is overclocking."

When I looked puzzled, Herbie explained, "Roy buys an old 90 MHz PC at a garage sale for 50 bucks, then puts in a faster clock chip and $700 of other parts and makes it go at 137 MHz. Running the microprocessor faster than it was originally designed to go is called overclocking."

My puzzled face got more puzzled. "And why do you want to do that? Just wait six months and a faster computer will come on the market. What can you do with a 137-MHz processor?"

"It's not what you can do with it, it's whether you can do it at all. It's a hobby! Anyhow, the processor heat goes up with speed. So to successfully overclock my next project, I'll need a better heat sink than it has now. So which works better: a gold heat sink or a platinum heat sink?"

I croaked, "You use *platinum* in a hobby?"

Roy showed me an ad in the July issue of *Popular Manics*. It offered a whole range of after-market heat sinks for overclocking. The top of the line, "for better heat dissipation, is plated in 24K gold." In a highlighted box was the 24K platinum-plated "Millennium Edition Heat Sink—Run Your Processor for a Thousand Years!"

I flipped through the rest of the overclocking hobby magazine. In several other places there were ads touting souped-up mother boards and graphics cards sporting "gold heat sinks for better heat transfer."

I began, "Here's the simple answer. There's an acronym for this, spelled S-C-A-M." Then I tossed the magazine in the trash can.

Herbie chuckled and Roy protested, but I silenced him by uncapping a whiteboard marker dramatically.

"Welcome to Heat Transfer 101," I said, and began the following lesson:

Heat energy moves around in three different ways: **conduction, convection** and **radiation.** All three come into play in a heat sink, so we can use it as an example of how plating might make a difference.

Conduction (see Figure 3-18) is the flow of heat through a solid body from a hot spot to a cold spot. Heat enters the base of our heat sink from the relatively hot, overworked processor, and spreads through

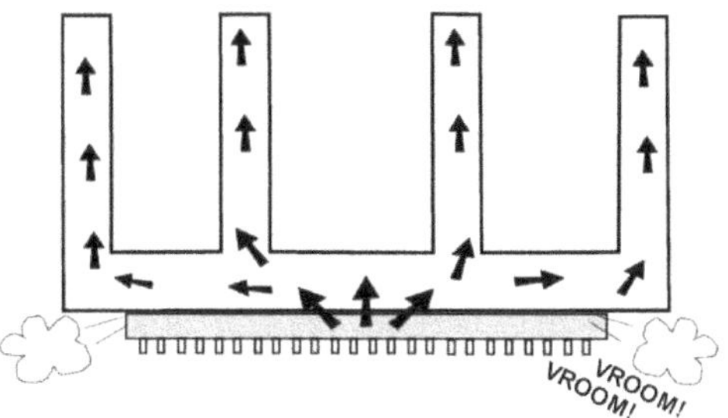

Figure 3-18

the body of the sink until it reaches the relatively cold surfaces touching the air.

Some materials are better heat conductors than others. A good conductor, like aluminum, will have a small temperature difference from the hot spot to the cold spot, which is good, because it means the processor temperature will be lower. But what if your typical aluminum heat sink is plated with gold?

Gold is about twice as good a conductor as aluminum. That's good. But plating is very thin, and makes up less than 1% of the whole heat sink. The improvement in overall conductivity would be less than 1%,

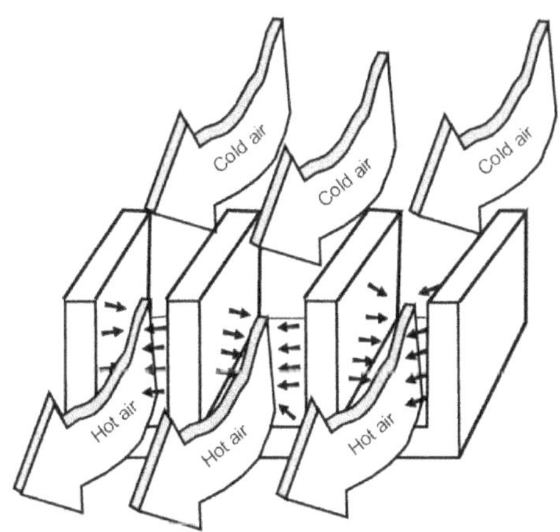

Figure 3-19

which is negligible. It's like trying to reduce a 1K resistance by putting a 100K resistor across it.

Now platinum is only half as good as aluminum, so it will actually reduce the heat sink conductivity by a similarly negligible amount.

Conclusion: Gold or platinum plating would have no effect on conduction, unless it were so thick that you couldn't afford it. A solid platinum heat sink would actually be worse than a solid aluminum one.

Convection is the flow of heat from the surface of a solid body into a fluid (liquid or gas) moving by (see Figure 3-19). Heat *convects* from the surface of our heat sink fins into the surrounding air. The air can be moving for any reason (wind, fan or buoyancy of the heated air). The temperature difference for convection depends on the shape of the surface and how fast the air is moving. But it doesn't depend at all on what the solid is made of. It could be wood, plastic, sandstone or peanut butter. For a given amount of heat, the temperature difference between the air and the surface of the fins would be the same.

Conclusion: gold or platinum plating has no effect on convection.

Radiation. Heat can flow away from any surface by infrared radiation (see Figure 3-20). That's right—you are giving off electromagnetic radiation right now, since your skin is most likely warmer than the room you are inhabiting. Our heat sink is probably hotter than

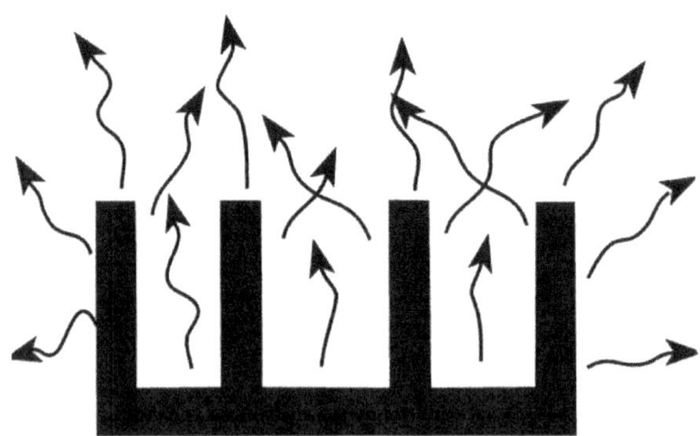

Figure 3-20

the box it inhabits, so it can give off heat by radiation. The amount of radiation it can emit is limited by a surface property called *emissivity* (oddly enough). A surface that emits the theoretical maximum infrared radiation for its temperature has emissivity of 1. Any real surface has emissivity between 0 and 1.

Now here's a property that plating can really change! A bare, brand-new, clean and shiny aluminum heat sink has emissivity of about 0.1, which is not very good. But exposed to ordinary air for only a few days, aluminum oxidizes and starts to look like a bathtub faucet covered with soap scum. That increases the emissivity to about 0.6, which is much better.

But plate the aluminum in pretty, shiny, non-corroding gold or platinum, and the emissivity goes down to the neighborhood of 0.01. That's right—the "thermal enhancement" actually *prevents* radiation, making the heat sink and the processor hotter. Cheaper finishes like black anodizing, gold chromate or even paint can easily give emissivity of 0.9, if you want to maximize radiation.

Table 3-2 summarizes the benefits of adding precious metal plating to a heat sink.

Roy looked sad, as if he had just heard that Santa Claus was a fake. "Wow, I never heard anything like this before. What magazine did you get it from?"

"*Heat Transfer* by J. P. Holman," I said, handing him my copy of the college textbook. Roy opened it to the middle and saw an integral equation. He put the book down gingerly and retrieved his magazine from my trash can.

"The gold looks nice, but the platinum is a limited-edition collector's item. I think I know which one to get now," he said, and zipped away.

Table 3-2
Scorecard for Gold or Platinum Plating

Conduction	Negligible effect
Convection	No effect whatsoever
Radiation	90 to 98% worse

Herbie smirked at me. "Did you, at least, learn anything?" I asked.

"Yeah. Save the gold and platinum plating for where it really belongs—body piercing jewelry," he said with a smile.

IMPROVING THE WEAKEST PLAYER

Alphonse sat in his office, hunched over a schematic and a half-assembled circuit board, behind stacks of magazines on sports and the stock market. "Sorry I missed your big meeting. A small emergency at the environmental lab. Herbie got trapped in my web of thermocouple wire growing out of the HBU rack," I said.

"Tragic!" Alphonse said. "Were you able to salvage the thermocouples?"

"The Rescue Squad got there ahead of me," I said.

"We all have our priorities," he said, "Too bad you couldn't see the big sales presentation. Manufacturing and Regulatory were there, and they gave me the go-ahead. Now all I need is a quick sign-off from you and we can get this project back on track."

The project currently off-track was the Neural Net Echo Canceler. It is supposed to smooth things out when Herbie's HBU gets a nervous tic. The main stumbling block, in my previously ignored opinion, was poor air flow. The average air velocity across the electronic components was a pitifully small 100 feet per minute.

Alphonse handed me a thick sheaf of color presentation slides. "Here are the notes from the Therm-O-Goop sales guy. They've got this great new printed circuit board material with really fantastic thermal properties. Not only is it going to make my board work thermally, but I have a feeling it will revolutionize the printed circuit industry. I just went online and bought a big chunk of Therm-O-Goop stock."

I flipped through the pretty slides. Most were glossy photos of circuit boards held by attractive models of all genders wearing lab coats and safety glasses. "Just what is this wonder stuff?" I asked.

"Exsulator Hi-K. It's a dielectric material that replaces the epoxy/glass layers in a standard circuit board. Its dielectric properties are only so-so. The thing that you'll be most interested in is here…" he said, and flipped to nearly the last page of the pile of slides, "…the thermal conductivity of Exsulator Hi-K is more than 10 times higher than standard epoxy/glass" (see Table 3-3).

"Impressive—for an insulator," I said, "but how does this help your Neural Net Echo Canceler?"

Alphonse looked at me incredulously. "Help me how? You know what's wrong with my board (see Figure 3-21). It's covered edge-to-edge with 36 signal processor chips in BGA packages. They get rid of their heat mainly by conducting it into the board. And most of the BGAs are just too darn hot. If I change the material from epoxy/glass to Exsulator, that will make the thermal escape route 10 times better. That should make the component temperatures drop like a loogie into the Grand Canyon, shouldn't it? The beauty of this is, I can fix the thermal problem without making a single change to my schematic, bill of materials or artwork!"

I scratched my head and said, "That would be nice, wouldn't it? Sounds almost too good to be true."

"You sound suspicious," Alphonse said. "Me, too. That's why I said it's only 10 times better. Therm-O-Goop claims it's almost 20 times

Table 3-3

Dielectric Material	Thermal Conductivity (watt/meter °C)
Epoxy/glass	0.2
Exsulator Hi-K	3.5

Figure 3-21 Alphonse's board: How well can a printed circuit board spread heat if it is completely tiled with high power components? Even if it were 100% copper?

better. So I discounted their hype by 50%. Ten times improvement should be plenty to suck those high temperatures down."

"You would think," I said, "but I have a feeling the improvement in thermal conductivity is going to be a little less than 10 times."

"How much less? I can be as skeptical as the next guy. Is it only eight times higher? Seven?" he offered.

I said, "Let's be generous and say that the Therm-O-Goop sales rep was being honest. I'll concede that Exsulator Hi-K is 17 times more conductive than epoxy/glass. What is important is not the conductivity of the dielectric material, but the conductivity of the whole printed circuit board. Don't forget, there are a lot of copper traces and vias in that board, too. What we really want to compare is the thermal conductivity of the board made of copper and epoxy/glass against a board made out of copper and Exsulator Hi-K."

"How do you do that?" Alphonse asked.

"To get an exact value is harder than parallel parking an SUV without hearing glass break. But to get an approximation is actually quite easy. So easy that it is usually covered in a heat transfer textbook in

Chapter One. If the copper is fairly evenly distributed throughout the board, then you just average together the thermal conductivity of the copper and the dielectric, based on their percent of volume of the board. I think your board, with multiple power and signal layers, and lots of vias, qualifies.

"I like to use the letter k for conductivity. So we could estimate the total conductivity for a board with this equation:

$$k_{pcb} = \%_{copper} \text{ X } k_{copper} + \%_{dielectric} \text{ X } k_{dielectric} \quad \text{Eq. (3-9)}$$

"Let's plug in some numbers. Copper has a very high thermal conductivity—about 380 watt/meter/°C, but there isn't a lot of it in a circuit board by volume. Copper makes up about 2 to 3% of the volume in a typical board. That gives us:

$$\text{Epoxy/glass: } k_{pcb} = 0.03 \text{ X } 380 + 0.97 \text{ X } 0.20$$
$$= 11.6 \text{ W/m °C} \quad \text{Eq. (3-10)}$$

$$\text{Exsulator Hi-K: } k_{pcb} = 0.03 \text{ X } 380 + 0.97 \text{ X } 3.5$$
$$= 14.8 \text{ W/m °C} \quad \text{Eq. (3-11)}$$

Alphonse looked disappointed (see Table 3-4). "We improve the material that makes up 97% of the board by 17 times, and the total improvement to the thermal path is only 28%? It should be 1,700%! How can that be?"

I said, "Compared to copper, epoxy/glass is a horrible conductor of heat. So horrible, that you'd have to improve it a lot more than 10 or 20 times to make a difference. Let me put it this way—what if

Table 3-4

Dielectric Material	Effective k_{pcb}
Typical epoxy/glass $k = 0.20$	11.6
Exsulator Hi-K $k = 3.5$	14.8
Improvement in total k_{pcb}	28%

I doubled my scoring in basketball? Would you finally let me on your Lunchtime League team?"

Alphonse had no trouble making that calculation in his head. "No! Seeing as you score about one point per game, doubling that would not exactly make you into the next Michael Jordan. And don't give me that routine about you being a defensive specialist."

We argued the relative merits of defensive skill and scoring ability for the rest of the afternoon. Eventually Alphonse realized that even 28% was *some* improvement in conductivity. So I ran that through the Therminator to see how much the component temperatures would go down if we changed the material to Exsulator Hi-K.

With epoxy/glass, the highest component was about 86°C above ambient, which was about 36°C too hot. The miracle dielectric Exsulator Hi-K reduced that component by a whopping 2°C.

"Are we going to increase the air flow now?" I asked after showing Alphonse the results.

"Don't bother me about that now!" he said, tapping his keyboard furiously. "I've got to dump that Exsulator stock before the market closes!"

Getting Lost in the Cracks

"Why are you being such a Mother Superior about this heat sink mounting?" Herbie complained.

"A mother-what?" I retorted.

"Remember the old TV show *The Flying Nun?* Mother Superior was the head of the convent that was always saying no to the Flying Nun, whenever she wanted to entertain some orphans or spend the weekend on some wealthy casino owner's yacht," he explained. "That's who you're being now."

"You're the one who tried to sneak a 27-watt component in place of a 5-watt component," I responded. "Maybe you deserve to get whacked across the knuckles with a wooden ruler."

"OK," he said, "don't get your habit in a bunch. So I tweaked the power up a little! I'm giving you all the real estate you want for a bigger heat sink. Isn't that penance enough?"

"How generous!" I said. "Sure a bigger heat sink helps. But I also need some holes in the board for a spring-loaded mounting bracket."

"That's the problem, Mother Superior. My component is a BGA that has 1,296 leads that need to get routed away from the package to the rest of the board, and you want to drill giant holes through the board, smack in the middle of my fan-out pattern. Just so you can attach your fancy-shmancy, spring-loaded monument to thermal design!" he said. "Why can't we just glue the darn heat sink on, like we've done a hundred times before on other boards?" he said.

I thought for a second about how to put it in Herbie lingo. I said, "When you crank up the component power by an order of magnitude like this, you have to start paying attention to a lot of details you ignore when the power is low. It's like the different levels of cleaning your house. When your kid brother stops by to watch TV, do you bother to clean up anything before he gets there?"

Herbie scoffed, "For Mervin? No way!"

I went on, "When your girlfriend is coming over, maybe you take the pizza boxes off the couch, so she can sit down without getting a stain on her clothes."

Herbie added, "And I kick all the dirty socks and underwear on the floor into one neat pile."

"But what about that time she invited her father to dinner at your house?" I said. "And you cleaned the place like you were going to sell it? You even rewired all the light switches so they pointed up when the lights are on."

"The guy is a church minister and former Marine Corps drill sergeant," he said.

"So when the pressure is on, you have to get all the details right," I said. "The same thing applies to heat sink attachment. Only instead of high-profile houseguests, I'm dealing with a temperature budget.

"According to your component data sheet, the operating limit is 100°C case temperature. The air coming into the shelf will be 50°C. That gives me a temperature budget of 100–50 = 50°C to work with. There are a bunch of causes of temperature rise that I have to fit into that 50°C budget." I couldn't think of how to say the bullets in a bullet list out loud, so I scribbled them out on the whiteboard.

Temperature rise budget. Temperature increase due to:

- Heat from all the upstream components
- Convection thermal resistance between the air and the surface of the heat sink

- Conduction or spreading resistance between the surface of the heat sink and its base
- Thermal resistance in the crack between the base of the heat sink and the case of the component

After capping the smelly marker I went on, "I can deal with the first three items by locating the heat sink in a good spot on the board, making the heat sink as big as possible and making the heat sink out of a good conductor like aluminum. Those are the parts of heat sink design I deal with first, the pizza boxes on the couch, so to speak. The crack between the sink and case is more like the dust behind the TV, something we could ignore when power levels were low.

"But at 27 watts per component, it's time to break out the dust rag. If we glue the heat sink to the top of the BGA, even using thermally conductive glue, the thermal resistance of that joint, when done perfectly, in this case might be about 1°C/watt. When the component power was 5 watts, that meant a 5°C rise between the component and the heat sink. Big deal. At 27 watts, that gives a 27°C rise between the component case and the heat sink. You have eaten up more than half my temperature budget in this one detail that I'm supposed to be able to ignore!"

"Oh," Herbie said in a tone of suddenly finding a Tenth Commandment he had forgotten about. "That sounds bad. But couldn't you just make the heat sink a little bigger?"

I said, "I might be able to make up for the temperature rise in the glue joint by making the heat sink bigger. But in this case, I would need four to five times the surface area, which could make the sink bigger than the whole board. That seems undesirable to me."

Herbie said, "Ouch. Bigger than the whole board? I suppose you'd still want more than a couple of mounting holes in the board for that, too."

"You bet your wings, Sister Bertrille," I said.

"So how does your spring-loaded gizmo fix the problem?" he asked.

"I would like to get the thermal resistance between the top of the component and the heat sink down to zero. Maybe we could put the heat sink in a furnace, then jam it onto the BGA so it melts onto it like cheese onto a hamburger. But that method has some obvious disadvantages" (see Figure 3-22).

"The next best thing I know of is to polish the bottom of the heat sink so it is nice and flat and smooth, then put a thin coating of thermal grease on it and smoosh it down on top of the part. The grease squeezes out of the way in the places where the microscopic high spots on the heat sink can touch the high spots on the component case, but fills the tiny gaps. That gives the minimum practical contact resistance, maybe about 0.1°C/watt," I explained.

"But the factory people hate grease. After a day of greasing heat sinks, they wear it home on their best T-shirts," Herbie said.

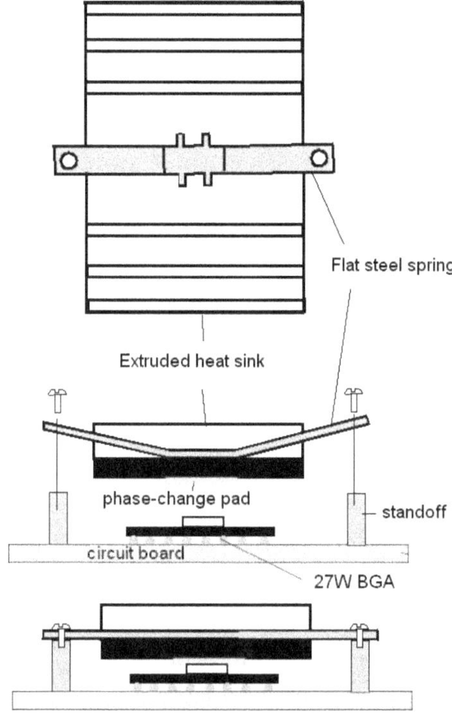

Figure 3-22 When component power goes up, the joint between the heat sink and component package gets to be more important. You don't want to lose your whole temperature budget in that crack.

"I know," I said, "And grease doesn't act like glue. It acts more like—well—grease. It prevents things from sticking together, not the other way around. So you need something else to keep the heat sink from wandering away, like a mounting bracket.

"But now there's 'greaseless' grease. They call it phase change material, which is a fancy way of saying that it melts. It's solid at room temperature, so it can be placed on a heat sink as a thin pad. When the component turns on and heats up, the pad softens into a grease-like goo, and the thermal joint is formed. But that means that when the heat sink is put on and the mounting bracket is screwed down, the pad is still cold and solid. The heat sink has to be able to squish closer to the component when the pad melts, so that means the mounting bracket has to be spring-loaded. When the pad melts, the spring force smooshes the heat sink into the goo and holds it there. That's why I need the mounting holes for my fancy-shmancy spring-loaded heat sink mounting bracket."

"All right already!" Herbie said, "Maybe I can work in your mounting holes by moving the TeleLeap logo to another place in the board silkscreen."

"What?! You were going to compromise the thermal design of the module for a *decoration*?" I fumed.

Herbie shrugged, "You may be as strict as Mother Superior, but those branding guys think they're God."

Section 4

Radiation: No, Stefan and Boltzman Were Not a '70s German Heavy Metal Band!

There is one good thing about thermal radiation. As the ambient increases, and the component temperatures go up with it, heat transfer by radiation actually increases. There is more radiation at an ambient of 50°C than at an ambient of 20°C, even when the temperature difference is exactly the same. The radiation is bigger: Not a whole lot, but nothing to sneeze at, as my allergic wife likes to say.

That is about the only good thing about radiation. It is hard to predict mathematically in complex electronic assemblies, it is nearly impossible to measure and when it is supposed to work in your favor, you can't count on it.

But infrared cameras are really fun toys to play with in the lab.

SEEING (INFRA)RED

When Herbie says his girlfriend Vernita "really lights up a room," it is literally true (see Figure 4-1).

Not just for her, but for everybody and even everything that exists. True, that is, if you can see infrared (or thermal) radiation.

Vernita lights up a room because her body glows like a lightbulb— heat rays shooting out in all directions. The amount of heat she radiates goes up with her skin temperature and her surface area, according to this formula:

$$\textbf{Radiation} = \sigma\varepsilon AT^4 \qquad\qquad \text{Eq. (4-1)}$$

σ is the Stefan-Boltzman constant, another property of the universe brought to you by the law firm of Stefan and Boltzman, equal to 5.669×10^{-8} W/m^2/K^4, ε is the emissivity of her surface (how well she rates on a scale of 0 to 1 against a theoretically perfect radiator), A is the surface area of her body and T is her surface temperature, in absolute (Kelvin) degrees.

Absolute temperature is how many degrees you are above absolute zero, which instead of zero is really –273°C. Human skin temperature

Figure 4-1 Vernita lights up a room, infraredly.

is around 30°C, so her absolute skin temperature is 273 + 30 = 303K. The next time you take your kid's temperature, tell her or him it's 300 degrees.

Vernita's surface area is about 30 square feet (2.8 square meters) and her ε is maybe 0.9. Punch all that into a calculator and you get about 1,100 watts.

No wonder she lights up a room. She's carrying a 1,000-watt heat lamp around with her (see Figure 4-2).

If that doesn't seem realistic to you, pat yourself on the back: 1,000 watts is about 900 calories per hour. Vernita could lose weight by standing in an empty room and metabolizing like crazy just to maintain her body temperature. That doesn't happen. Here's why.

The walls of the room, being at room temperature (by definition), also radiate. So Vernita, while shooting out her 1,100 watts, is also being showered with radiation from all directions by the surrounding walls. Who wins or loses this radiation contest depends only on the temperature of her skin and the temperature of the walls. The heat she loses or gains by radiation is given by this equation:

$$\textbf{Net Radiation} = \sigma\varepsilon A(T_{\text{skin}}^4 - T_{\text{walls}}^4) \qquad \text{Eq. (4-2)}$$

Figure 4-2 The walls of the room radiate back nearly as much.

Assuming the walls are at a normal room temperature of 25°C (298K), then the net radiation she gives off to the room is only 30 watts. That still lights up the room, but not enough by which to read this chapter.

Now you know more about thermal radiation than my buddy Lester at BGAs R Us, Inc. I wanted to make a thermal model of one of his BGA packages in the Therminator, so I could predict its thermal performance on a new circuit board. So I stole the details for the BGA's internal construction and thermal properties from a paper that Lester wrote.

When the computer model of the BGA was done, it needed to be checked out against experimental results. Fortunately, Lester's paper provided that, too. He had measured the thermal resistance ($\theta_{j\text{-a}}$) for the BGA in "still air." I could run a Therminator simulation matching the conditions of his experiment, using my new model of the BGA. If my results matched Lester's experimental results, I could pat myself on the back and conclude that my BGA model was accurate. Then I could use it in other simulations and trust the results.

Lester's test was a fairly typical method for measuring $\theta_{j\text{-}a}$. The BGA was soldered to a 4×4-inch circuit board, mounted horizontally inside a wind tunnel with the wind turned off. While applying 5 watts to the die, he measured the air temperature and the die temperature. Dividing the temperature rise of the die above the air by the power gave $\theta_{j\text{-}a}$. Simple.

It was simple to model his experiment, too. Except that instead of Lester's 12.2°C/W, the Therminator came up with a $\theta_{j\text{-}a}$ of 21.0°C/W. Whoa! Computational fluid dynamics (CFD) may not be perfect, but it isn't *that* imperfect.

The big difference happened because of one assumption I had made in building my model. I had ignored radiation. In CFD it is easier to ignore radiation than to include it, so I got in the habit of neglecting it. Lester's paper didn't say anything about radiation one way or another.

So I called him on the phone. (I used to think it was hard to get through to people who publish technical papers. In reality, it is pretty easy. They publish because they want people to know what they did, and they are usually disappointed when nobody ever asks them about their work.) "But of course radiation is included in my measurement of $\theta_{j\text{-}a}$!" Lester admitted. "The heat comes out of the BGA by convection to the air and radiation to the walls of the wind tunnel. I deal in reality. You can't just turn radiation off in real life like you can in a simulation" (see Figure 4-3).

True enough. So I turned on the radiation option in the Therminator, reran the simulation, and got a value of $\theta_{j\text{-}a}$ of 13.2°C/W. That was within 8% of Lester's experimental result, so I concluded that my BGA model was pretty darn good.

So why should you care about the radiation from Lester's BGA? The whole point of this story is to give you yet another reason to *not* put much stock in the published value of $\theta_{j\text{-}a}$ from the component manufacturer. The CFD simulation of Lester's estimated that the heat lost by radiation was 64% of the total. And that 64% is included in his value of $\theta_{j\text{-}a}$.

So what? Who cares whether the heat gets out by radiation or convection or conduction, as long as it gets out?

Here is why you need to care. In Lester's test, the walls of the wind tunnel were at room temperature, so there is a lot of radiation (64%)

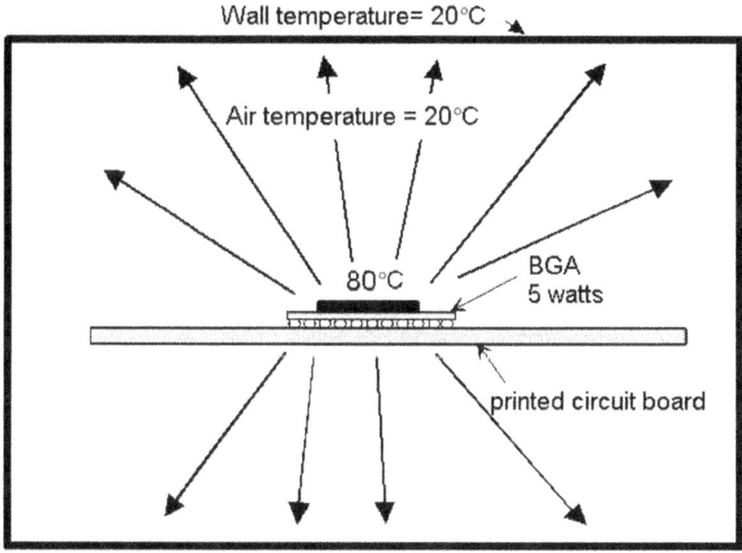

Figure 4-3 In Lester's test setup the BGA can radiate to the cooler walls.

lost from the BGA to the walls. But unless you are designing a circuit board to be used inside Lester's wind tunnel, the situation your board will see is probably more like this (Figure 4-4):

There are circuit boards on both sides of yours, with components at least as hot as the ones on yours. Remember the equation for Vernita's net radiation? The net depends on the difference in temperatures. If the component is nearly the same temperature as the "walls" that surround it, then the net radiation will be practically zero. It might even be negative.

Lester's value of $\theta_{j\text{-}a}$ depends on an important heat loss path that you just can't count on being there in your application. To give you an idea of how important this might be, in Lester's wind tunnel, where the BGA can radiate heat away to room temperature walls, the junction temperature is only 81°C. In the simulation, with radiation turned off, the junction temperature is 125°C, even though the surrounding air temperature is exactly the same (20°C).

This is just one example. I wish I could tell you that the radiation loss is always 64% of the total, and then you could factor that into

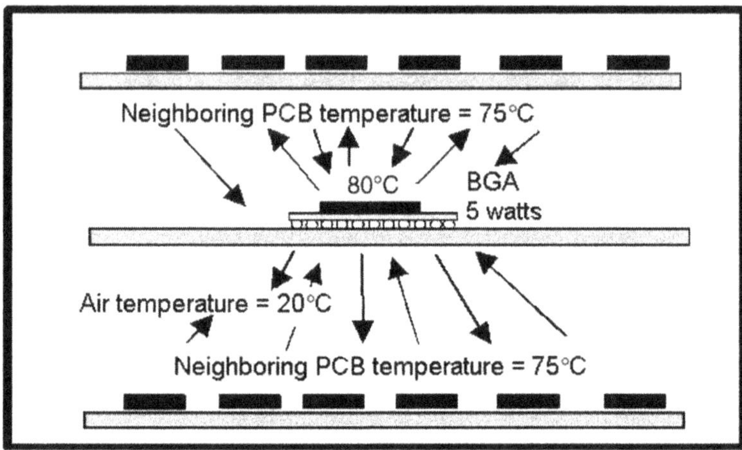

Figure 4-4 In your design the neighboring hot boards radiate back almost as much as the BGA radiates out.

your calculations. But that percentage differs with package type, air flow rate, power dissipation and the amount of copper in the circuit board.

The only recommendation I can safely make about using $\theta_{j\text{-}a}$ is: *Don't.*

And don't worry, Herb. Vernita still lights up a room when she comes in. It just works the other way around when she goes out in the sun.

THE BEAUTY OF IR IS ONLY SKIN DEEP

H erbie loves high-tech toys.

When I was unpacking my new infrared (IR) camera he hung around, handing me tools, flipping through the manual and generally slobbering like a beagle at breakfast time. "Is it working? Huh?" Herbie panted, "Can we try it out on—um, somebody?"

"Somebody? I thought you wanted to scan a circuit board for hot spots," I said.

"Sure, later," he said, "But I heard about a certain camcorder that had an infrared feature, but they had to recall it, because—er…"

"Yeah," I said, rolling my eyes, "it supposedly could see through people's clothes."

"Wait!" he backpedaled, "I just thought it would be cool, or even useful, if the camera could see through, uh, things that, like, block your view."

"First," I lectured, "It's against the thermal analysts' code of ethics to use thermal technology to spy on anybody's naughty bits. The company would probably bounce you for doing it, too. And second, it doesn't work. I already tried."

"But what about the recall they always talk about in the Internet news groups?" Herbie said. "And Robocop can see people through brick walls with his thermal vision!"

"Robocop is science *fiction*," I said, "The camcorder story is only *exaggeration*. Infrared radiation, which is how an IR camera senses the temperature of objects, is not like X-rays. IR is wimpy. It can't penetrate a single sheet of paper. Even window glass is opaque to infrared. The camera can only make an image of the outside surface of an object. I think somebody used the camcorder infrared feature to video a person wearing loose-fitting clothing. The clothing was warmer where it actually touched the person's body, and cooler where it was hanging loose, so on the screen you could see a vague outline of the person's body, instead of the shape of the clothing. That gave them the impression it could see through clothes."

"But even an outline —"

"Take a look at this IR image of yourself (Figure 4-5). Even if you scanned somebody who was naked with the IR camera, you would see less detail than you would using your (naked) eyes! That's

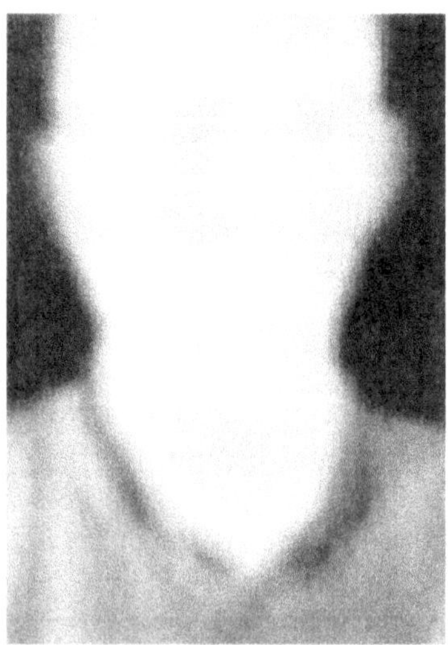

Figure 4-5 Infrared brings out all of Herbie's best features.

because the colors only indicate surface temperature. Because your skin is all the same temperature, your face is a fuzzy outline of all one color. In this picture we can't even tell if your eyes are open or closed, because your eyeballs are the same temperature as your face. Not very useful for satisfying your curiosity about hidden temperatures, is it?"

Things an IR Camera CAN'T Do

1. It Can't See Plumes of Hot Air.

It would be great if we had a camera that could show heat flowing from place to place. Herbie's manager commanded me to bring over the IR camera so he could see the hot air coming out of the vent of Herbie's new shelf. I had to explain to him that air is about the only thing that is transparent to infrared, so the camera *can't* show you hot air. The explanation wasn't good enough. He had seen *Robocop,* too. So I did a quick demo for him using the monitor of his desktop computer. The image showed the hot surface, and some hot components inside the cooling vents, but no plume of hot air coming out. He could see his warm hand in the image hovering over the monitor grill, but no hot air. He could even feel the hot air with his hand, but couldn't see it on the image. That was enough to finally convince him. He seemed more disappointed than he should have been. It turned out what he really wanted was to borrow the camera to look for air leaks around the window frames of his old Victorian-style house.

If you think about it, if air weren't transparent to infrared, the camera wouldn't be of much use. All you would see in the image would be the layer of air right in front of the camera lens.

2. It Can't See Copper Traces on the Inner Layers of a Multilayer Circuit Board.

Herbie did have a valid reason for wanting to see through things. He was worried that some of the traces on the inner layers of his latest board might be a little thin for the current they would be carrying. Thin traces carrying lots of current heat up from their own electrical resistance, and that heat should make them visible. Unfortunately, the camera can only see the outer surface of the circuit board. By the time the high temperature of an inner trace reaches the outer surface, it has also spread in all other directions, so if you see any hot spot at all, it is likely to be an indistinct blob. That may be useful, but it won't look like an X-ray image that shows exactly where the electron bottleneck is.

3. It Can't See Though the Sides of Shelves, Cabinet Doors, Brick Walls or Component Packages.

Are you getting it yet? Infrared radiation doesn't go through anything, except air. OK, here are some exceptions—specially made slices of silicon, which can act as IR lenses and windows, and very thin films of plastic.

4. It Can't Measure Temperature Accurately.

The latest IR cameras are digital. They either interface to a PC and store the infrared image and temperature readings digitally, or store the same information internally, to be read by a computer later (see Figure 4-6). This is quite useful because long after the image is taken, you can open the picture on the computer and use your cursor to display the temperature at a particular location. You don't have to squint your eyes and try to match up the color in the image to a Color/Temperature scale

This gives the impression that with one snap of the camera, you can measure all the component temperatures and toss away those messy thermocouples. In my dreams! As luck would have it, the reading from the IR camera is most likely *wrong*, for these reasons:

- For the camera to see the components, the board has to be removed from its shelf or chassis, or at least the cover has to be removed. That means:
    - ➤ The air flow around the board is not the same as in its real operating environment.
    - ➤ Heat from neighboring boards is missing.

Figure 4-6 A typical computer display of an IR image of a circuit board in operation. (Managers are much more impressed when the images are in color. Note: the very dark parts on the board are either very cold or have low emissivity surfaces. It seems unlikely they would be colder than the surrounding board, even if they dissipate no heat. Don't be fooled by a low temperature measured by an IR camera.)

> Many components are not dissipating their true power, because the board cannot be completely exercised outside the chassis.

- Nobody knows the *emissivity* of the components. The radiation given off by a surface depends on two things–its temperature, which the camera is trying to measure, and the emissivity of the surface. Emissivity ranges from 0, for a very poor emitter, to 1, for a perfect emitter. The camera assumes all the surfaces have emissivity of 1, which is not necessarily true. It measures the radiation coming from a surface, and using the assumed value of emissivity, calculates a temperature. If the assumption is wrong, the temperature is wrong. It is hard to know the emissivity accurately, because it depends on the material, its microscopic roughness, corrosion, cleanliness and other things, some of which change with time.

So just because the display tells you the temperature is 67.2°C, don't think you know it to the tenth of a degree. Probably more like ±10°.

What an IR Camera *Can* Do

The IR camera is useful, even with all its non-X-ray-type limitations. Even though its temperature readings may not be accurate, the camera can't be beat in finding the hot spots on a printed circuit board. It is great for finding surprises—an undersized fuse will glow like Rudolf's nose on Christmas Eve as soon as you turn on the power. It would take hours or days to probe each and every component on a board with thermocouples, and nobody has the patience to do it (at least I don't). Sometimes the surprise is that a part I expect to be hot turns out to be just a smidge above room temperature.

So in a snap, the IR camera can record the hot spots on a board. I use this to help me decide where to place my thermocouple probes, so that with the board properly mounted in its shelf or chassis, with all the correct covers closed and the right signals passing through all the channels, I can measure the real operating temperature of the components. Without that first pass with the camera, some unexpectedly hot components might sneak through my selection process.

The IR camera should be in the toolbox of anybody doing electronics cooling. Like any tool, it has its uses and abuses. Like CFD it gives very pretty color pictures. For some reason color pictures trick people into thinking the data is more accurate and reliable than it really is. Don't fall into that trap.

And if you see somebody snooping around your work area with an IR camera, don't worry—unless you see Herbie wearing those X-ray specs he recently ordered from the back of an *Archie* comic book.

"NEGATIVE RESULT—VERY IMPORTANT, TOO!"

"Do I got to read this whole thing?" Herbie asked, flipping through my latest thermal test report. "Can't you just tell me the answer in five words or less?"

"Check the so-called executive summary. It's only two words," I said.

"OK—'Executive summary: needs fixing.' That's it?" he said. "Needs fixing? What needs fixing?"

"That's why there are 11 more pages," I answered.

Herbie read further, his lips moving. Clearly he was not happy with my negative report. Then he went back on the offensive.

"I just noticed," he said, "you got a nice color infrared picture of my board in here."

I said, "I used that to identify the hot spots on your board, to make sure I didn't miss anything when I attached my thermocouples for the thermal test."

"I seem to recall," he said in a tone reminiscent of Perry Mason cross-examining Lt. Tragg, "another report with a color temperature picture in it. Now I remember: I had you do a thermal simulation of my board, so we could predict the component temperatures to make sure they'd be OK ahead of time."

"That's right. The Therminator gave a nice color picture of the board, with colors representing temperature, a lot like an infrared photograph," I admitted. For some reason I began to perspire.

Herbie smiled. "How come, then, you don't put both of those pictures, side-by-side, in the thermal report here, so we can see how good a job you did at predicting?" he asked.

That was an amazingly good question, I had to admit. TeleLeap paid pretty big bucks for The Therminator to predict temperatures, and fairly large bucks for an infrared camera to measure temperatures, and me a reasonable amount of bucks to operate both of them. Perhaps Herbie and my other thermal customers had a perfectly good right to see a figure like Figure 4-7. If he did, it would usually look more like Figure 4-8.

Hesitantly, I pulled out my original Therminator analysis report on Herbie's board. I hesitated, suspecting this comparison wasn't going to be pretty.

"Nice job," he gloated, placing the two reports on the table in front of me.

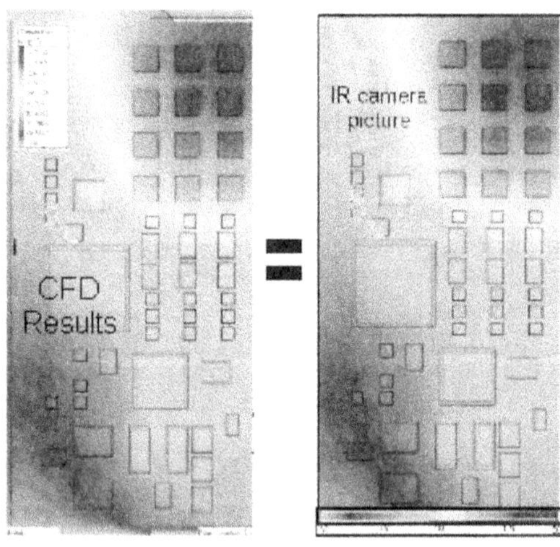

Figure 4-7 A figure one might expect to see at the end of a thermal analysis report, in some perfect world.

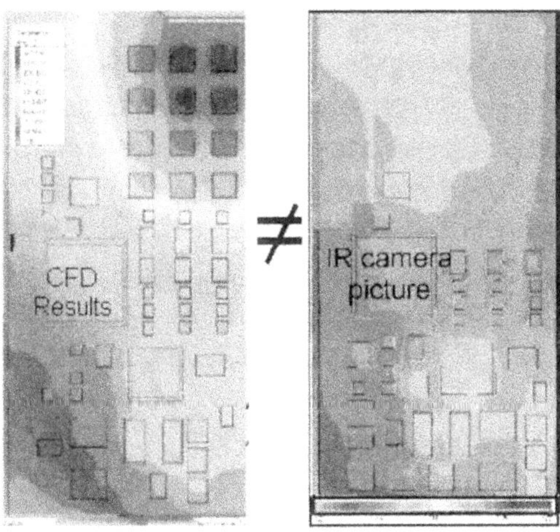

Figure 4-8 What you really get, 9 times out of 10. Except you'll never even see this anywhere.

After a long silence, during which I pretended to intently study the two pictures as if for the first time, I announced, "This is what we in the thermal sciences call a *negative result.*"

I learned about negative results early in my career, while working on my Senior Design Project for my bachelor's degree in mechanical engineering at the University of Detroit back in the latter half of the 20th century. My semester-long project was to design a magneto-hydro-dynamic (MHD) energy storage device, an idea frequently proposed by my physics teacher. If you think a liquid mercury flywheel sounds like a nutty way of storing electrical energy, then you are way ahead of where I was. My advisor, Professor Jimmy Chu, approved the project, and a mere two hours of hand calculations later, I proved that the concept could never work. Too much energy lost in viscous friction. But that wasn't the worst part of the problem. What the heck would I work on for the rest of the semester?

"Negative result—very important, too!" Professor Chu told me, and that phrase has stuck in my heart ever since. He accepted my two hours of work as credit for the whole semester, explaining that my negative result would save everybody else from wasting time on that concept, including that physics teacher. He said you could often learn more from negative results than from success.

So instead of "losing" Herbie's reports in the back of my filing cabinet, I was inspired to dig into my files for more examples. Turns out that 9 times out of 10, comparing the Therminator pictures with the infrared camera images ends up with a negative result, and the one time there is some agreement, it is just a coincidence. It was something I had long suspected, but just never admitted to myself.

Jimmy Chu was right. There were lessons to be learned by looking at those pictures side-by-side. And after looking at enough of them, I began to get some ideas for improving the thermal simulation process (maybe.)

Lesson 1 (or Excuse 1 in Herbie's parlance): Musical Components.

Take another look at the pictures that don't match (Figure 4-8). One reason they differ is that the components are *not* in the same place! It's common practice for the designer to keep making changes to the board design after I have finished the thermal simulation, many times without letting me know. Not only does it make it hard to match up the simulation and infrared pictures, but it could mean that my temperature prediction no longer holds water. So the lesson is: If you're going to change the board design after the thermal simulation, at least run that change past the thermal guy.

Lesson 2: Power Garbage In, Temperature Garbage Out.

Nobody seems to be good at estimating the power of electronic components. I know I'm not. So I rely on the circuit designer to figure it out. But they tend to overestimate, because they have been taught that to use the maximum power for everything in their calculations is the conservative way to design. That may be good for sizing power supplies and trace widths, but it makes the temperature predictions come out too high. The higher the power estimate, the higher the temperature prediction, no two ways about it. In Herbie's example, the power estimate for the whole board in the Therminator was 16 watts. After the board was built the total power was measured at 6 watts. One memory chip was estimated at 5.5 watts. In the flesh it dissipated 0.1 watt. Don't be amazed if the IR camera can't spot a chip dissipating 0.1 watt.

Lesson 3: One Picture Is Worth a Thousand Lies.

An infrared camera has a really hard time getting a picture of a circuit board working in its real environment (see Chapter 4.1). Usually a board is stuck in a box, surrounded by other boards and, covered with cables, and closed up behind a couple of solid metal doors. That's the environment that the Therminator is trying to duplicate. To get an infrared picture, I usually shoot it while it is plugged into an open test fixture on a bench in the lab. The air flow conditions are nothing like the real environment. Maybe the only time you should expect the CFD picture to match the infrared picture is if you are simulating the test fixture (and even then Lessons 1 and 2 will mess things up).

Lesson 4: Don't Hide Negative Results.

I think the Therminator does a pretty good job predicting component temperature. I spend a lot of time learning its intricacies, making sure I know how to use the various turbulence models, when to apply temperature-dependent thermal conductivity and how much detail to use in calculating radiation losses. By confronting "negative results" face-to-face, I realize that all that stuff is small potatoes, considering I can't get the components to stop shifting around or even get power estimates within 50% of their real values.

That's why you'll never see any pictures like Figures 4-7 and 4-8 in thermal test reports, or even in technical journals. It is nearly impossible to get pictures like that to agree, and nobody likes to publish their "unsuccessful" results, especially in full color. The absence of those figures ought to remind you what Professor Chu taught me.

"Negative result—very important, too!"

SELECTIVE SURFACES

"Who's that guy with the gigantic novelty scissors?" I asked.

"That's Percy," Herbie said, "VP of the Biological Intercommunications Group, BIG for short. The Lost Pet Tracking System has been his baby since day one."

There was a festive air at the Prairieton Police Station—kids holding dogs on leashes, colorful balloons tied to squad cars, even the high school marching band. Herbie had invited me to the first full-scale field demo of the Lost Pet Tracking System (LPTS).

Percy, using a police bullhorn while standing on the tailgate of an Animal Control pickup truck, continued his speech, "LPTS is powerful, yet simple. It has three basic parts. One is the ID chip that is implanted under your dog or cat's skin. The second part is an array of Remote Units in strategic locations all over town. They scan for ID chips, and when they detect the one you are looking for, its location is reported back to the third part, the PetMap Display back here at police headquarters. Your police dispatcher then directs an officer to the lost pet in real time to within a 3-foot radius. Result: no more lost puppies! I'd call that a real triumph for technology."

Together Percy and the mayor awkwardly wielded the gigantic scissors against the fat ribbon stretched across the police station driveway. A photographer from the weekly paper snapped their picture while the band played "How Much Is That Doggy in the Window?" After hacking through the ribbon, Percy smiled and announced, "For years police have used dogs to hunt down fugitives. At last they have a way to find the dogs when they become fugitives themselves."

The field demo was essentially a high tech game of Hide and Seek. Five dogs with ID chips were sent out (with masters in tow) to find hiding places. After 10 minutes, the police dispatcher, using LPTS, would attempt to guide police cruisers to find them. The public was invited to follow the progress on the PetMap Display on a big-screen TV that had been specially set up on the front steps of the courthouse.

At a nod from Percy, the police chief fired his gun in the air to start the demo. After a flurry of barking and rear-end sniffing, the dogs set off in all directions.

"While we're waiting for them to hide," Herbie said, "I actually have a thermal-type question for you."

I almost said, "Shoot," but the chief still had his gun out, so I just said, "OK."

"It's about the Remote Units," he said, "They don't generate much heat themselves, but they are mounted in small, unventilated boxes outdoors, on light poles and such, where the sun can beat down on them. We were worried that the solar heating might cause the electronics inside to get too hot."

"That's worth worrying about," I said.

"The actual pole-mount boxes were designed and built for us by another company—PoleCat Enterprises. We make the electronics and stuff them in. When I mentioned my concern about solar radiation, the PoleCat guys told me they could fix that real easy, using what they called a *selective surface* coating on the outside of the box," he said.

"So they painted it white," I said.

"Right! How did you know? Seems to work. But what the heck is a selective surface?" he asked.

"Ah, selective surface," I said. "High technology figured out by the Bedouins, or some other desert dwellers, a few thousand years ago. It takes a little bit of physics to explain, but not too much."

I drew this graph (Figure 4-9) with a stick in a bare patch of dirt in the police lawn.

"Every object, unless it's absolute zero, emits thermal radiation," I explained. "If you graph the amount of radiation a particular object puts out as a function of wavelength, it looks like a teepee. It emits a little energy at every wavelength, but most of the energy is in a fairly narrow wavelength band. The location of this teepee-shaped band depends on the temperature of the object. The higher the temperature, the smaller the wavelength.

"For example, the sun, at 6000°C, is on the left of the wavelength graph, in the range where you have visible light. This teepee in the middle could be a hunk of iron in a blast furnace at 500°C, just barely glowing red. And over here on the right is your Remote Unit box, at maybe 100°C, not glowing at all in the visible spectrum, but still emitting some infrared radiation at much longer wavelengths."

Herbie nodded, "High temp, low wavelength, I got it."

"That's the transmitting side of radiation," I continued. "Now let's talk about the receiving side. Most surfaces, like this dark blue shirt I unfortunately chose to wear on this extremely sunny, hot, July after-

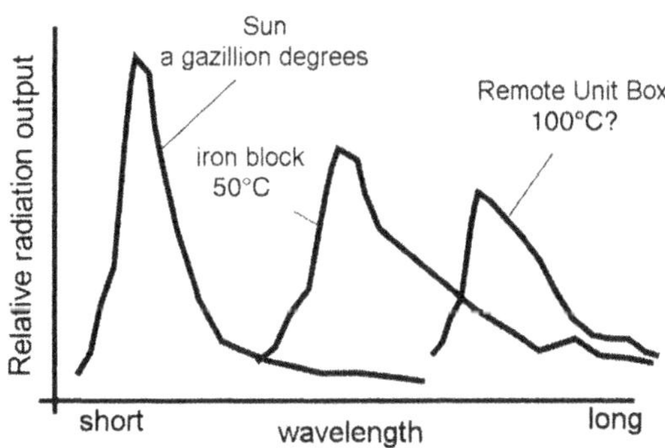

Figure 4-9 Hot things (like the sun) radiate at shorter wavelengths than cold things.

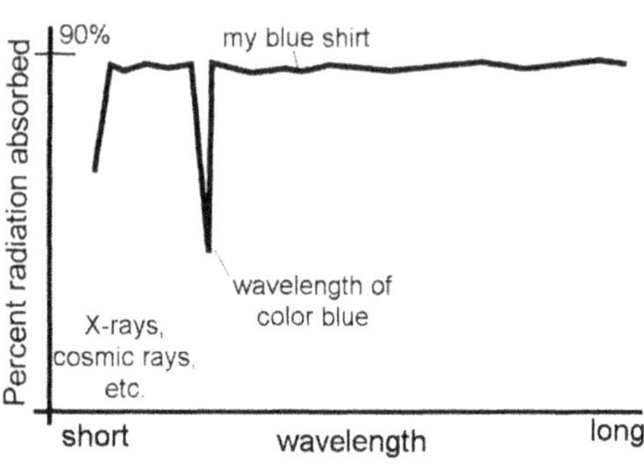

Figure 4-10

noon in the northern hemisphere, absorb radiation equally well across the whole spectrum of radiation, except the real far end stuff like X-rays and cosmic rays. The absorption graph for my shirt would look like this" (Figure 4-10).

"A selective surface gets its name because it absorbs radiation only at *selected* wavelengths (see Figure 4-11). Here is what the absorption graph for your special white paint probably looks like. In the wavelength region of solar radiation, it absorbs very little. Most of the short wavelengths, especially visible light, get reflected. At longer wavelengths, it absorbs normally. If I had on a white shirt like that, I wouldn't be sweating as much as I am."

Herbie smiled, "Wow. It's like a high-pass filter for radiation! Great idea. But we don't need to absorb anything. Don't they have a surface that is a bad absorber at all wavelengths?"

"Sure," I said, "It's called a mirror. That would work, too. But if you don't keep it cleaned and polished, over time the dirt and gunk on the surface turn it back into a pretty good absorber".

Murmuring from the crowd drew our attention back to the Hide and Seek demo. Something was wrong. A couple of the blips on the big screen kept winking out.

Herbie winced and immediately saw the problem. "Looks like the Remote Unit in Grid B-3 is going intermittent," he muttered. "Come on. Let's go. Maybe a squirrel is chewing on the antenna lead again."

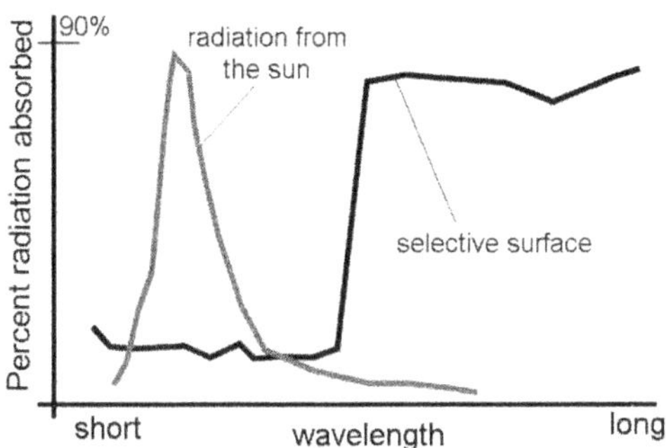

Figure 4-11 A selective surface absorbs radiation only "where the sun don't shine."

We hopped on our company-issue scooters and puttered toward the northwest corner of Prairieton. Herbie skidded to a stop in front of the Elks Hall near the grain elevator, and pulled out a heavily creased map. He cast his eyes around, looking up at the rooflines of the surrounding commercial buildings. "Now where the heck is that Remote Unit?" he said aloud. "They usually put them on a city light pole, so they can wire power from the light circuit."

"Is that it?" I asked, pointing to the FoodMax grocery store across the street. There was a box about the size of a Stephen King hardcover novel mounted on the south-facing cinder block wall, about 10 feet above the pavement.

"Yeah, I couldn't recognize it at first," Herbie said, "and now I know why. I guess FoodMax didn't think the Remote Unit went with its *decor.*"

Somebody had painted the Remote Unit box a deep chocolate brown, to match the color of the rest of the building. As we sat there on our puttering scooters, one of the Demo dogs trotted by the store and lifted his leg against the wall directly under the Remote Unit. He seemed to smile slyly at us, his tongue hanging out in the hot July sunshine.

I shrugged and said, "I guess if you're going to depend on a selective surface for temperature control, you have to be more selective what surface you mount it on."

This chapter first appeared in the November 2002 Issue of *Electronics Cooling.* It is reprinted with permission.

Section 5
Tales of the JEDEC Knight

There are two mistakes in cooling electronics that are more widely practiced than any other. Both are based on outdated industry standards, and they are equally dangerous. It would be hard to say which one is more popular—ask your colleagues and you will probably find that they subscribe to both of these myths.

One is that you can double the life (or reliability) of an electronic component for every 10°C you reduce its temperature. Read Chapter 30 of *Hot Air Rises* for why that is just flat out not true.

The second is the incredibly popular practice of using θ_{ja} and θ_{jc}, component thermal resistances to find junction temperature. That's just not right. Even the industry standard that defines these resistances tells you not to use them for that purpose. The following chapters chronicle my sporadic campaign to slap the wrists of those thermal engineers whom I find doing it, and tell them to cut it out.

*CIRCUIT BOARD *NOT* INCLUDED

Herbie and Renatta were arguing so loudly that I was sure it had to be about the football pool. Renatta always picks the team with the best passing quarterback, while Herb bets on the team geographically closer to his home town.

"But they'll make Swiss cheese out of it!" Herbie protested, I assumed, about what the Lions would do to Green Bay's front line.

"Mmmm, Swiss cheese," I intoned *à la* Homer Simpson.

"There's the guy who can settle this," Renatta said, pointing to me. She had been furiously sketching on the whiteboard, and Dry Erase ink flecked her cheeks.

"Oh, football's not my game," I said.

"Football? Get your mind back to the office, big guy," Herbie said. "We're talking strictly thermal here."

"Phooey," I said. "I was hoping to hear something interesting. What's this drawing with all the X's and O's? "

"That's my new concept for saving the telepathic port module (TPM) from thermal disaster," Renatta announced, muscling in front of Herbie, "We'll use the circuit board itself as a heat sink to cool the psychic/optical transformer (POX)."

Renatta was a mechanical designer visiting from the Sedona lab. It was her unenviable task to package the HBU, putting "the 4.54 kg of brains into a 2.27-kg box," as she put it. Renatta's father had been the Carter administration official in charge of converting the United States to the metric system.

"Ever since she read your book," Herbie said, "Renatta thinks she's a junior thermal guru."

"So you're the one!" I said. "Why don't you tell me about your heat sink idea?"

Renatta gave me the forced smile of a toll both operator. "I got this idea from a thermal chat room on the Internet. This guy Spike says that if you put copper planes in your board, you can cut the thermal resistance of the component in half!"

Herbie nodded in agreement, "Your *Thermal Tip of the Day #142* even says that a lot of the heat from a component flows into the board through the leads, so that makes sense."

"OK," I said tentatively, "So what are the X's and O's?"

Renatta said, "This other guy named KirkNSpock said he read somewhere in a magazine that you can improve the heat sinking ability of the board even more by drilling a whole bunch of holes in it. 'Increases the surface area', or something, he said. So that's what I want to do to the TPM—add a whole bunch of holes right around the POX. And I need Herb to tell me where it's OK to add holes without cutting traces and stuff."

"Holes in the board?" I asked. "I never heard of that before. How do they help get rid of heat from components?"

"I wondered about that, too," Herbie said. "But since that anonymous Internet guy read it in a magazine, it has to be true. He said he didn't know how it worked either, but claimed that tests showed that holes made the components cooler."

"My theory," Renatta said, "is that air flows through these holes, creating turbulence, which increases the rate of heat transfer. Plus you expose the edges of all the hot power planes to fresh air, which makes the whole board cooler, too. That makes me think the holes should be pretty big, about 10 to 25 mm in diameter."

"What's that in American money?" Herbie asked.

"About the size of a penny," I said cents-ibly.

"My idea is that the holes increase radiation from the board," Herbie. " I saw in *Scientific American* at the barbershop last month that a cavity can act as a blackbody radiator. We want to cover the board with mini-blackbody cavity radiators, so they should to be small holes, about 1/16th-inch diameter."

"Wow," I said. "That's not the kind of magazine my barber carries."

Renatta said, "So we need you to settle the dispute. We know we want to add a bunch of holes to this board to make it a better heat sink. Tell us how big and where they should go to give us the best thermal performance."

I stared at the board layout for a minute, then pronounced, "To get the absolute minimum temperature rise on the board, you only need one hole, about 1/8-inch diameter, right here by the input fuse."

Herbie looked at the fat dot I had drawn on the layout. "But that will cut right through the trace that supplies all the power to the board!" he figured out.

"Exactly," I said. "No power, no heat, no temperature rise. That's the only way drilling holes in a circuit board can reduce its temperature."

"But the increased surface area…" Renatta tried.

"Anything bigger than a pinhole throws away more surface area than you gain," I said.

"But what about increased turbulence…" she said.

"Unless you turn the board perpendicular to the fan, there won't be any flow through those holes, so they won't have any effect on turbulence," I explained.

"But the cavity radiation…" Herbie sputtered.

"For radiation to be important, there has to be another surface in line of sight that is cooler. What can your holes radiate to—the neighboring board, which is at the same temperature? Sounds like your common sense radiated out of that cavity in your skull."

"Hey!" Herbie whined, his hand unconsciously feeling around the top of his head for an opening.

I said, "Let's talk about using a printed circuit board as a heat sink for a component. Guess what? You don't have any choice! Renatta, you mentioned the component thermal resistance. Do you know how they measure the so-called thermal resistance between the junction and the ambient ($\theta_{j\text{-}a}$)? The component is soldered onto a test board about 4 inches square. So that resistance value in the data sheet already includes the heat sinking effect of a fairly large board. When they show some fantastic new electronic toy in a TV commercial, the announcer always says at the very end *'Batteries not Included.'* I think on the data sheet next to the value of $\theta_{j\text{-}a}$ they ought to be forced to print * *'Standard test board not included.'*

"That's what I mean about you not having any choice. To have a thermal resistance at least as good as the data sheet, you *have to* use a board that is at least as good a heat sink as the one in their test."

Renatta sat stunned for a moment, then asked, "How do we know if our board is a better heat sink than theirs?"

"Good question," I said. "The answer is you can't, because they don't tell you about their test board. It might be a single-sided board with just a few signal traces. Sometimes it is a double-sided board with just signal traces. Sometimes it has power and ground planes inside. The more copper the board has, the better it spreads heat. How does that compare to your board?"

Herbie said, "We have 16 layers, with 4 power planes and 4 ground planes. Sounds like it should be better."

"Could be," I said, "or maybe not. After all, you have more than just the one component on your board. If you have lots of high-power stuff on the board, there are no empty areas for the copper to spread the heat to."

Herbie chewed on his lip. "There are 12 POX chips on the board and 16 other high-power chips," he said.

The lab became very quiet.

Renatta started to erase the whiteboard. "You don't like the holes idea just because it came from a chat room," she said.

"Wait, wait!" I said. The ink smearing, filling the holes, had given me an idea. "I think KirkNSpock may actually be right!"

Herbie said, "I knew the cavity radiation…"

"No, this has nothing to do with radiation, or air flowing through the holes. When that Internet guy said you could reduce component temperature by adding holes, he didn't mean *holes;* he meant that other kind of hole, the kind you use all the time on the board to connect layers together—what do you call them?"

"Vias?" Herbie said.

"Exactly!" I said, "Vias! Of course! Tiny holes that go through the board, lined with copper! That would make the board a better heat spreader. One problem with using a board as a heat sink is that most components are surface mount. The leads only touch the signal layer on the top of the board, which has hardly any copper in it. You really want to connect to those big, fat copper planes on the inside of the board, but the heat can't get there, because it is separated from them by layers of insulating epoxy. But if you sink a whole bunch of vias through the board, you connect the top layer to the interior planes with copper, and the heat should spread a lot better. Your Internet guy was right—adding lots of vias can make the board a better heat sink. He just used a somewhat imprecise term that got you thinking down the wrong track. All vias are holes, but not all holes are vias."

"Vias! That's what I was going to suggest next!" Renatta said.

"Keep erasing," Herbie said, "Doesn't matter even if they do help. There's just as much no room for lots of vias as there was for regular holes."

"As long as it's settled, then," I said. "I'll go look for the football pool."

THERMAL I/O

Sillycon Valley—I^2R Industries announced its revolutionary, high-density DBGA (double-sided ball grid array) package. The increase in I/O density is achieved by a set of solder balls on top of the package, in addition to the full array on the bottom. "Customers screamed for lead density, and we responded with a vengeance," said Chip Schotz, Vice President of Packaging, "Now it's up to our customers to figure out how to connect to all those extra solder balls on top" (see Figure 5-1).

Don't do a spit-take with your coffee—there is no double-sided BGA. It would never see the light of day, simply because no one would ever buy it. Sure the density is unbeatable, but there is no practical method of soldering all the leads on such a package to a printed circuit board. No electrical hardware design engineer would accept such a ridiculous package. Then why do thermal engineers?

Interconnection Technology has been so well worked out over the last 50 years that we now take it for granted. It is hard to believe, but until the 1960s, electronic assemblies were made of discrete components bolted to slabs of phenolic and connected with point-to-point

Figure 5-1 The solder side and the *other* solder side of the new DBGA package contain plenty of I/O for today's high-density electronics.

("rat's nest)" wiring. Look inside that antique television in your uncle's basement and you'll see what I mean.

One benefit of the Cold War was that electronics had to be shrunk to fit inside missiles, while increasing reliability. This forced the development of a truly integrated system of interconnecting electronics. Printed wire boards eliminated point-to-point wiring. Component packages were standardized to be machine-soldered onto them. The boards plug together with separable connectors, and the only wires left are bundled together in connectorized cables. From the silicon chip to the system level, every piece is designed so that the electrical connections can be made simply, quickly, reliably and, most of the time, by machines.

When you are building 50 million DVD players a year, you have a strong motivation to want to solder all the components together in one pass. That's why no component vendor is going to introduce a package that can't be assembled to a printed wire board with readily available soldering technology. So putting balls on top of a BGA is a bad idea, unless you also have a simple, reliable, cheap and automated method of connecting those balls to the rest of the circuit. (If you do think of such a method, remember me at patent time.)

To get a comprehensive system of electrical interconnection takes a lot of effort to define all the input/output (I/O) paths at every interface. Component vendors think long and hard about electrical I/O before they release a new package.

Why don't they do the same thing with the thermal I/O?

Mainly because hardly anybody thinks that there needs to be a well-defined path for heat to get from the chip to the outside world. Giving it a name should at least get people thinking about it. And giving it an electronic-sounding name like "thermal I/O" will get much more attention than "heat transfer path."

Every electrical signal (plus power and ground) has a well-defined, dedicated path to get from the chip to the lead to the solder pad on a printed wire board. The heat generated on that chip needs to have paths from the chip to the outside world that are just as well-defined as the electrical I/O.

Heat finds paths out of the chip whether they are defined or not—through the leads, through the plastic overmold, through air pockets and decorative labels if it has to. The problem is, if those paths are not designed in and spelled out, then nobody knows where or how to connect them to a heat sink, or whether a heat sink will do any good at all. It's like getting a 596-ball BGA with no I/O table in the data sheet. You have to guess which signal is on each ball, if any.

Component manufacturers release 30-, 50-, even 100-watt parts willy-nilly. Often their data sheets don't have even a hint that they are aware that an electronic packaging engineer will have to attach some kind of heat transfer gizmo (a heat sink, jet-impingement nozzle or Freeze Ray courtesy of Batman's nemesis, Mr. Freeze) to carry those watts away. Years ago thermal I/O could be ignored. Power was low, and almost any sneak path for heat was good enough.

But 100-watt chips are not right around the corner. They are standing on the porch peeking in the front window. So I am invoking my power as a thermal guru to decree the following standard: Any vendor not complying with it will be severely made fun of in future publications (is that a scary enough threat?).

Thermal I/O Standard

1. Thou shalt clearly define thermal I/O paths from the chip to the outside of the package. The thermal resistance of each path

from the chip to its specified end point on the package surface shall be measured and published.

2. Thou shalt talk to customers and find out how they take heat away from boards, then design package thermal I/O paths that work together with those methods. No more tossing out new package designs and shrugging your shoulders when asked about cooling.

3. Thou shalt add to each chip an on-board temperature sensor that can be read externally on dedicated leads while the chip is in operation. That way a customer can tell if a cooling method is actually benefiting the chip inside.

4. Thou shalt give the operating temperature limit in terms of the junction temperature, measured by the sensor mandated above, so we have a chance of telling whether your part is getting too hot in our application. No more "70°C ambient" parts.

Perhaps the concept of thermal I/O will catch on, making our job of cooling electronics a little more possible. I'm just afraid we'll see the double-sided BGA in production first.

"Thermal I/O" first appeared in the February 2002 Issue of "Electronics Cooling." It is reprinted with permission.

JEDEC STANDARD: STAKE IN THE GROUND OR STICK IN THE MUD?

There is a conference on just about any topic. Really, any. But even I was stunned to hear that there would be a technical conference on Thermal Issues in Electro-Organic Systems. With all my work on the HBU, it seemed to be right up my alley. It also didn't hurt that it was in Orlando, Florida, in the middle of October.

I looked forward eagerly to hobnobbing with other thermal gurus, swapping Herbie stories and stealing brain-cooling ideas from the competition. Little did I suspect that in that friendly venue I would encounter one of my pet peeves.

We had been settled into the darkened hotel conference room for several hours, nearly lulled into a stupor by the whirring of the projector. But with the next introduction I sat bolt upright. Marshall! I had failed to notice his name on the program. But there he was, already directing attention to his first slide with his beady, little, red laser pointer.

Marshall and I had tangled before. He was the Moriarity to my Sherlock Holmes, the Wellington to my Napoleon, the Mandark to my Dexter. His laboratory was bigger, his thermal skills perhaps sharper than mine. If only he had not turned to the Dark Side and become a JEDEC Knight.[1]

Marshall began his talk, innocently titled, "Optimizing the Thermal Resistance of a Glucose-powered Flip-Chip BGA Package." Before he even finished showing the outline slide, I jumped up, muttering, "He's doing it again! He's doing it again!"

All eyes turned to me and glared. Marshall cleared his throat, and I meekly sank down on my hard, hotel ballroom chair.

Marshall works for an electronic component manufacturer that was recently spun off and renamed Vaguetron, I think. He heads a staff of thermal engineers who work to improve the thermal performance of their products, which sounds great, in principle.

The paper was very straightforward. Marshall had looked at several construction parameters of the GPBGA to see how they affect its thermal resistance. He tried three die sizes, two die-attach adhesives, three plastic encapsulants, two substrate materials and three locations for the platinum glucose reservoir. He used CFD simulation to look at all the combinations of these parameters, and confirmed those simulations by testing prototypes in a wind tunnel. The audience bubbled with excitement at the graph showing good agreement between the CFD predictions and the wind tunnel data. Marshall proclaimed only two parameters were important for low thermal resistance: large die size and a high-conductivity substrate dielectric. The rest of the factors were not important. He also claimed that this result would dictate the final list of improvements to be made to the package design.

After a round of enthusiastic applause (Marshall is a dynamic speaker), I again jumped to my feet to ask a question. "I'll be good," I said, nodding to the program chair, who had risen from his seat.

Marshall pretended he didn't know me, smiling as he gestured for me to speak. "Very good work," I said, "but I don't understand one thing. Your goal is to make a package with the minimum thermal resistance, I assume for the purpose of having the lowest possible die temperature in the customer's application."

Marshall nodded, getting suspicious.

"Can you explain why you optimized it for a situation that is nothing like any real application? In all your tests the package is attached to a 4×4-inch board with no other components on it!" I continued.

Marshall flipped back to a picture of his experimental setup (see Figure 5-2). "Of course I can explain," he said, "This is the industry-standard JEDEC test board. It is the board used to evaluate the thermal resistance of all packages. Not only at Vaguetron, but at every component maker in the known universe." Then he quickly took a question from the other side of the room.

I stayed on my feet, hoping to follow up. Arguments raced through my head. It was driving me nuts that he was using an exquisitely powerful tool like CFD simulation to optimize the thermal design of components, but in such a way that they would never be optimum in any real application.

A component package has many different paths for heat to get from the die, the source of the heat, to the outside air. They can be divided into two main directions—down into the circuit board, and up to the top surface, where heat flows into the air, or into a heat sink, if there is one.

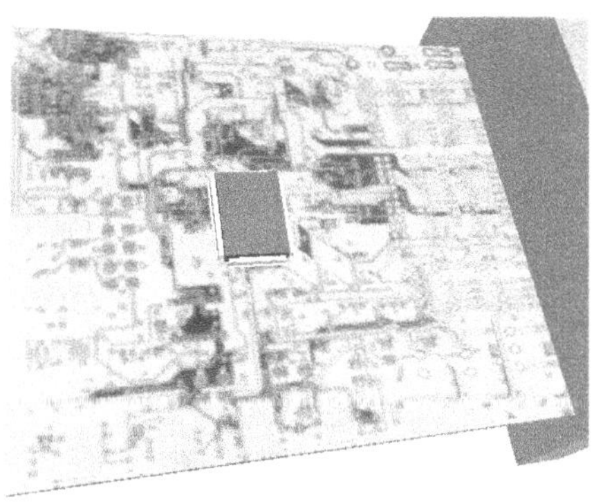

Figure 5-2 The JEDEC standard board for measuring thermal resistance of a component package. Does this look like the boards you design?

Following the JEDEC standard, Marshall put his component package on a circuit board about 4 inches square. Even the two-layer version of this board has enough copper in it that it spreads heat away from the bottom of the package. The JEDEC board makes a pretty good heat sink, considering that it has more than 60 times the surface area of the component alone.

"What's the problem?" I imagined Marshall retorting. *"To be used, any component must be attached to some kind of board."*

But when have you ever seen a real board where every component was allowed 32 square inches of its own real estate, counting both sides?

Having that monstrous heat sink attached to the bottom of the package exaggerates the importance of all the heat paths down to the board. Any improvements in the heat paths to the top of the package look dinky in comparison. So of course Marshall's test would show that a high conductivity substrate on the bottom has a big effect on die temperature, but the platinum glucose tank on top does practically nothing. If he had done his study with a 1×1-inch board on the bottom and a big heat sink on top, his conclusions might have come out completely opposite.

Finally, Marshall came back to me. I asked, "Why optimize the package for best thermal performance on the JEDEC board? Isn't it like optimizing the fuel economy of a car only for the stop-and- go driving of Manhattan? If that were the standard, we'd conclude that making cars more aerodynamic is useless."

Little arguments began to stir up in the audience. Marshall shrugged, "I grant you that the JEDEC board may not be representative of every obscure way that a component package might be applied, such as the unusual products *your* company makes. But considering that Vaguetron must design for all kinds of applications, what standard would you suggest?"

While I hesitated, open-mouthed and finger raised, the program chair announced a coffee break. He physically led me by the elbow to the sweets table and stuck a Danish in my mouth.

"Marshall won that round, but I know you're right," he said. "The JEDEC standard is pretty unrealistic and leads to packages that don't work as well as they could, but it's the only standard around. Until they come up with a better one, you'll never get Marshall, or other component vendors, to change."

I chewed my Danish in silence, contemplating the day when the moribund $\theta_{j\text{-}a}$ Empire would crumble. Perhaps, I imagined, if I could truly master the power of The Farce against him, Marshall could be turned…

Notes

[1] JEDEC is a real, actually benevolent industry standards organization called Joint Electron Devices Engineering Council. It struggles valiantly to update the thermal resistance standards to make them more useful.

Section 6

A Collection of Not Even Loosely Related Stories

For some reason they decided we needed to learn Set Theory in the fourth grade. It drove me nuts. Not that the rules of Set Theory were hard to memorize. Intersection and Union, big deal. What bothered me was the definition of a set—there wasn't any definition! A set could be anything, even nothing (remember the null set?). In my fourth grade mind (which hasn't changed all that much yet), a set was supposed to have something in it, and those things were supposed to be related to each other somehow. A salt and pepper shaker set. The set of all even numbers. There is even the set of George Bush and Bill Clinton (left-handed U.S. presidents.)

But according to Set Theory, there can be The Set of All Objects that Have No Relationship to Each Other. The paradox of that definition drove me crazy to the point of failing my Set Theory test. Because if the thing that defines the membership of the set is that the members have no relationship, then that definition is something they all have in common, so they *do* have a relationship, which means they can't be members of the set. So the following set of chapters can't logically exist. Hurry and read them before they collapse in on themselves in a spiral of self-contradiction.

THE MILK-BOX PROBLEM

Another thing that keeps me awake at night is the Dairy Products Delivery Specialist (milkman). Milk is delivered to my house straight from the local dairy. They deliver it in the early, early morning—about 4 a.m. The dairy has a policy that when the outside temperature is below –18°C (0°F), the milkman rings the doorbell. That warns me that the milk is sitting on the front porch, and that it might freeze before I get up at a decent hour and fetch it in.

The last time I was awakened by that 4 a.m. doorbell, the second thought that came to my mind was, "How fast does the milk get cold anyway? Do I really have to get out of bed to rescue the milk, or can I wait until 7 or 8 a.m., when I am already up and dressed?" Luckily, I am just the person to answer that question for myself.

Here is the problem as you might see it in a heat transfer or milk delivery technician textbook (Figure 6-1):

The milkman delivers 2 gallons of milk to the front porch at 4 a.m. The milk is put in a milk box, a colorfully decorated plastic picnic cooler. It is the worst weather in weeks, air temperature –29°C (–20°F) and the wind is blowing 10 miles/hour. The milk comes out of the delivery truck at 5°C (41°F). How long after the doorbell is

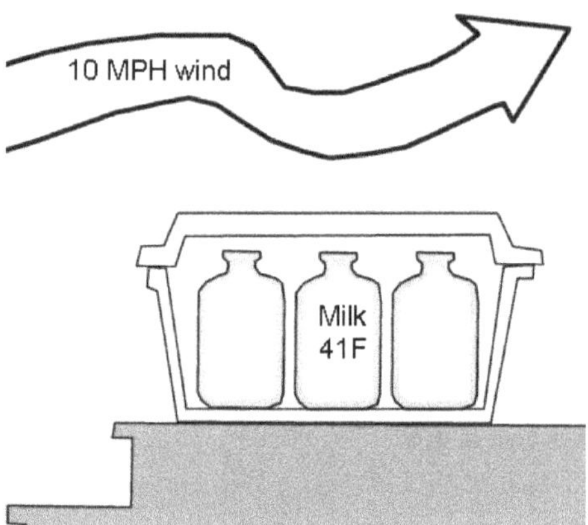

Figure 6-1

rung can I stay in bed before the milk cools down to 0°C (32°F) and starts to freeze? Freezing is bad, because it might burst the bottles, or even worse, damage the delightful dairy-fresh flavor.

This can be a tricky problem to solve in all of its detail, because I have introduced the idea of time. In most of our dealings with Herbie, we only cared about the steady-state temperature of electronics, that is, how hot it gets after a long period of time. We didn't talk about how fast this process of heating or cooling goes. That requires the idea of thermal capacity.

Thermal capacity is how good a substance is at storing heat. Strictly speaking, it is the amount of heat it takes to raise the temperature of a fixed amount of material by 1 degree. For example, it takes four times as much heat to raise the temperature of a kilogram of water by 1°C as it does a kilogram of air. You can say that water is four times better at storing heat, ounce for ounce, than air.

The milk, because of its thermal capacity, and because it is hotter than the outside air, has heat energy stored in it. That heat has to flow through several resistances in its path (see Figure 6-2), before it reaches the outside air. If we can quantify these resistances, and the heat energy stored in the milk, we can calculate how fast the temperature of the milk is going to change.

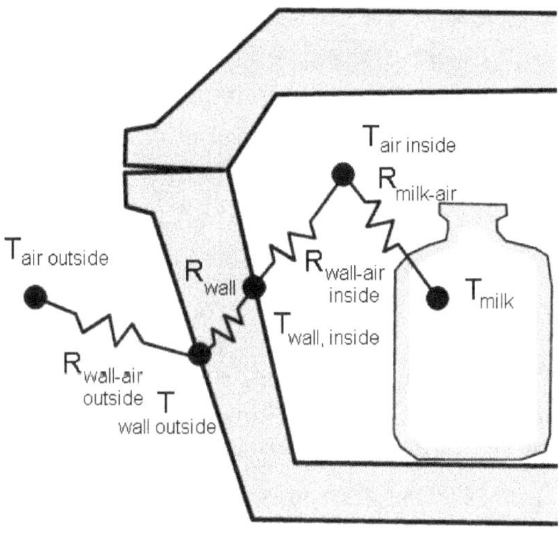

Figure 6-2

I have simplified this problem by assuming that the air inside the cooler is all at one temperature, and the milk is all one temperature. That way I can treat the heat flow as one-dimensional, and I can add all the thermal resistances together between the milk and the outside air. I think this is safe, since I only want to know the answer to the nearest half-hour, for sleeping-in purposes. I estimated the thermal resistance between a single bottle of milk and the outside air to be about 16°C/ watt. (See notes at the end of the chapter about how I estimated this value.)

Now I can write an equation that says the energy lost through heat transfer to the outside air equals the change in heat energy stored in the milk:

$$\Delta T_{milk}/\Delta \text{ time} = (T_{air, \, outside} - T_{milk})/(R_{total} \, mC) \quad \text{Eq. (6-1)}$$

m is the mass of the milk, and C is the thermal capacity of milk (about the same as water, 4,000 Joules/kg/°C).

The left-hand side of the equation is the speed at which the milk temperature will change, in °C/second. The right-hand side of the equation tells you that the speed depends on the temperature difference between the milk and the outside air. It will start cooling off fast, when the temperature difference is big, but then cool more and more slowly as the milk temperature gets closer and closer to the temperature of the outside air. The fastest cooling happens right at the beginning, just after the milkman drops the bottles in the milk box, because that is when the temperature difference between the milk and the air is the greatest.

I estimated the thermal resistance (R_{total}) between the milk and the outside air at 16°C/watt. If you put in some realistic numbers for the mass of the milk (m = 1.9 kg) and the thermal capacity of the milk (C = 4,000 Joule/kg°C), the outside air temperature of –29°C, and the initial milk temperature of 5°C, you come out with a rate of change of about 0.00027°C/sec, or about 1°C/hour. That gives me a minimum of about 5 hours before the milk goes from 5 to 0°C and starts to freeze. That is a thermal calculation I like, because it means I can sleep in until 9 a.m.

For extra credit, please figure out how much more I can sleep because of the latent heat of freezing of the milk. Don't be too shocked if you discover that the heat loss it takes to freeze milk is a lot more than it takes to just cool it down 5°C.

This type of analysis came in handy recently when Herbie's cousin Andre called me in for some advice. He works on FoneCable, which is a box about the size of a paperback book, that can hang on the outside of your house. It combines cable TV and phone signals somehow, so that you can now listen to your favorite TV show on any phone in the house.

Andre discovered during stress testing that the FoneCable circuit board can be sensitive to very fast temperature changes.

"It works perfectly at any temperature you want," Andre explained, "but if you change the temperature very fast, like they do in the environmental chamber, weird things start happening, from just turning itself off, to broadcasting your phone calls onto every TV set in the neighborhood. What I need to know is, how fast is a circuit board likely to change temperature when it is mounted in an outdoor box. Does the weather really change very quickly? And if so, doesn't the box shield it somewhat?"

The FoneCable problem is amazingly similar to my milk box, except for one major difference: the board generates its own heat (P_{board}). That adds another term to the milk equation:

$$\Delta T_{board}/\Delta \text{ time} = (T_{air, \text{ outside}} - T_{board})/(R_{total}\ mC) + P_{board}/mC \qquad \text{Eq. (6-2)}$$

The heat generated by the board (P_{board}) is 5 watts. Because of this the board is about 20°C hotter than the outside air at steady state. I estimated a thermal resistance (R_{total} = 4.0°C/watt) between the board and the outside air, the mass of the board (m = 0.17 kg) and its thermal capacity (C = 1,400 watt/kg °C).

According to *The Weather Almanac,* by Frank Bair, one of the most outstanding temperature changes on record happened on January 22, 1943, in Spearfish, South Dakota, when the temperature rose from –4F to +45F in 2 minutes. That's a good worst case.

Let's find the rate of temperature change of the board in this situation. The outside air temperature starts out at –20°C(–4°F), and the FoneCable board is 20°C hotter, or 0°C. Suddenly the outside air skyrockets to 7°C. What happens to the board? Plug in the values and you get an initial rate of change for the board of 1.7°C/minute or 102°C/hour.

The conclusion of the milk-box analysis is that as long as Andre's board can tolerate temperature changes of about 2°C/minute, he should be OK. That is about 10 times faster than my milk cools. The reason his rate of change was so high is mostly because the board has much lower mass and lower thermal capacity. If Andre has trouble, he could increase his thermal capacitance by filling up his box with something like milk, or he could try hooking it up so it could ring the doorbell and tell the customer to bring the FoneCable inside the house when it gets too windy out.

Thermal Resistance Estimates

For those who wonder where I got the estimates for thermal resistances between the milk and the outside air, I will give you my math here. If you find any guesses that seem too wild, remember that I am doing this in my head while I am in bed at 4 in the morning.

1. Resistance Between the Milk Bottle and the Air Inside the Milk Box ($R_{\text{milk-air}}$)

I am ignoring the resistance of the walls of the glass bottle, assuming that the resistance between the surface of the bottle and the internal air dominates. The thermal resistance between a surface and surrounding air is

$$R = 1/(hA) \qquad\qquad \text{Eq. (6-3)}$$

where h is the convective coefficient and A is the surface area. This is natural convection inside a closed box. The air is not very free to move around, and the temperature difference is small. Instead of trying to calculate a value from some textbook correlation, I will just assume the value is at the low end of the range for natural convection—about 1 Watt/meter^2/°C.

The surface area I figure in my head to be about 0.075 m^2 for one bottle of milk. That gives a thermal resistance between the milk and the air inside the milk box of about 13°C/watt.

2. Resistance From the Inside Air to the Inside Wall of the Milk Box ($R_{\text{wall-air, inside}}$)

I use the same formula, and the same value for h. But the surface area of the milk box is somewhat larger, about 0.5 m^2. That gives $R_{\text{wall-air, inside}}$ about 2.0 °C/watt.

3. Resistance Through the Milk-box Wall (R_{wall})

This is a conduction resistance through some kind of insulation. $R = t/(k\,A)$, where t is the wall thickness (about 0.025 m), A is the cross-sectional area (again, about 0.5 m^2) and k is the thermal conductivity of the insulation. I give it the benefit of the doubt and say it is a very good insulator, $k = 0.05$ watt/meter/°C. So R_{wall} is about 1.0 °C/watt.

4. Resistance From the Outside of the Milk-Box Wall to the Air ($R_{\text{wall-air, outside}}$)

Because the wind is blowing and the temperature difference is somewhat larger, I guess the value of h around 15 watt/m^2/°C. With $h=15$ and $A = 0.5$, that gives $R_{\text{wall-air, outside}} = 0.13$ °C/watt.

When you add all those resistances together, the total resistance between the milk and the outside air is about 16 °C/watt.

Notice one weird thing about my estimates (if they are right). The insulation in the milk-box walls is actually only about 6% of the resistance between the milk and the outside air. Most of the resistance comes from the low value of natural convection inside the box. The box could be made from aluminum and the milk would not suffer much.

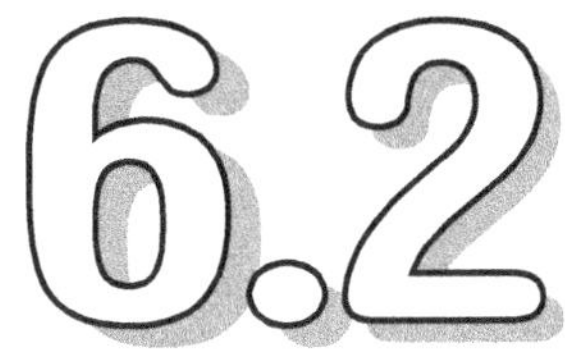

SPECS, LIES AND RED TAPE

Leon suggested that Herbie and I ride with him, and let the marketing and sales guys drive to the restaurant separately. At the first traffic light, the Mercedes full of suits went straight. Leon gunned his Geo Prism and turned right.

"Where you guys want to eat?" Leon asked. "Anywhere except where those guys are going. I'm buying—out of my own pocket. Someplace around here got burgers and beer?"

At the Lantern we hunkered over a small table of chili, cheeseburgers and a pitcher of beer. Herbie asked, "Why didn't you want to go with the rest of the guys from SpendTel?"

"Finish your beer and I'll tell you," he answered. Leon was a floor plan engineer for SpendTel. SpendTel's network planners purchase a ton of telecom equipment from dozens of suppliers. Leon's job was to figure out how to install all this stuff in the same building and hook it all together. He and a platoon of other SpendTel people were visiting TeleLeap to hear a spiel about our latest products.

Leon had looked out of place in the TeleLeap Product Demo Center, as did Herbie and I. My patched sweater, Herbie's jeans and Leon's golf shirt stood out among the crowd of shiny suits. Someone had requested that Herbie and I appear at the all-day meeting, and by now I was figuring it had been Leon doing the requesting.

"Here's why I need to talk to you guys alone," he began. "I need the straight dope on the UH-HUH 238."

"What's not straight? I think the customer literature on that shelf is pretty darn easy to figure out," Herbie said, getting a little defensive. He was quite proud of this product. The UH-HUH 238 allows a telephone company to offer a valuable new service to its residential customers. When a telemarketer calls, instead of hanging up on him, you push #238, and the call seamlessly switches to the UH-HUH 238. It keeps the telemarketer busy for up to 30 minutes by hooking him up to a digitized voice loop that says, "Uh-huh … yeah … I see … uh-huh…" While it does its dirty work, you can finish dinner.

"The 238 specs are easy," Leon said, "But that's the problem, they are too easy. Here is what they say: *'When using a fan to cool the U-238, the maximum inlet air temperature is 50°C, and the minimum air flow rate is 100 CFM.* That's great. But I need to put this shelf in the same cabinet with the TeleSell 590. That's the computer voice that calls you over and over trying to sell you vinyl siding."

"Don't they just cancel each other out?" I asked.

"Yeah, but not in two important ways," Leon said. "One is we make money on both services, and the other is the heat. The TeleSell 590 spec says it needs 300 CFM of cooling air at 50°C. So I plan to put in some fans that give 300 or more CFM, and then I want to stack your shelf right on top of it. But the problem is, the air coming out of the 590 is going to be hotter than 50°C."

"But you can't do that," Herbie said. "The air to our shelf can't be over 50°C. What's so hard to understand?"

"The hard part is that you're *lying.* The inlet air doesn't have to be limited to 50°C," Leon said.

Herbie turned red in the face. "We're not lying! Maybe there is some safety factor in that number, but it's based on actual test data!"

"You are, too, lying, just the same," Leon said calmly.

Herbie started sputtering, so I jumped in. "Yes, Herb, we are lying in the spec."

Leon filled up our beer glasses again and said, "We're all engineers here. We can talk reality. You know and I know that you can trade off inlet temperature for air flow. If the U-238 works fine at 100 CFM and 50°C, then at 300 CFM, it should be able to go to a higher inlet air temperature, like 60°C. What I need is for you guys to tell me what that trade-off is so I can engineer my cabinet correctly."

"I don't get it," Herbie said, "50°C is 50°C. We rate the U-238 to work at 50°C. It doesn't matter how much air flow their is. It's like the ratings on commercial grade components. They are rated to work from 0 to 70°C. It doesn't say anything in the data sheet about how fast the air is moving."

I shrugged my shoulders and said, "But the component vendors are lying, too. It's like the lie your parents told you when you were afraid of thunder—when they said it was just angels bowling in heaven. They didn't trust you to understand the real explanation, because it was complicated, and maybe they didn't quite understand it themselves. But Leon is right. You can trade off inlet air temperature for air flow. But putting that in a spec is complicated, and we don't trust our customers to understand it correctly. So we dumbed it down for them, and hoped it would still be useful."

Herbie asked, "So what should we have put in the spec? I thought you came up with the numbers yourself."

I started sketching on a Lantern paper place mat. "This is the story we'd have to tell," I said. "Inside the UH-HUH 238 are a whole bunch of components, but from a thermal point of view, you only have to worry about this one—the power supply brick. It is the hottest part in the shelf, and it has the lowest operating temperature limit. If this power brick is OK, then the rest of the U-238 is OK (see Figure 6-3)."

"Now, according to the Goodly Power Brick Co., this brick works only up to an ambient of 70°C. After I had a talk with one of their reliability engineers, and we found out we had gone to the same university and taken a class with the same jerk professor, he admitted to me the real temperature limit was 85°C, measured on the built-in heat sink. So, the goal is to make sure the heat sink of the power brick stays lower than 85°C.

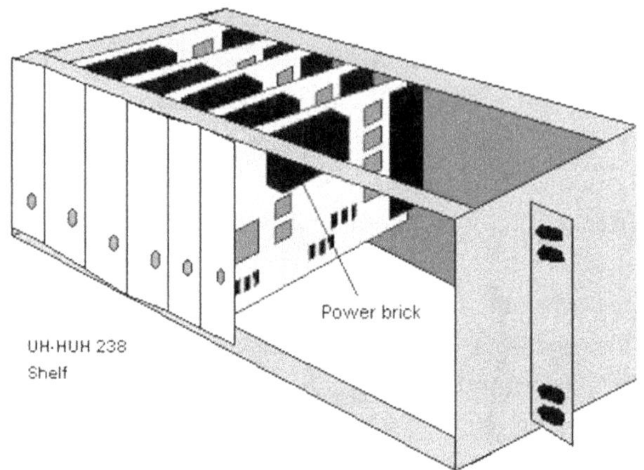

Figure 6-3

"We sell this product as a stand-alone shelf, so we don't provide the cooling system for it. Customers like you, Leon, mix and match it with all kinds of other equipment in the field. Some use natural convection, and some choose to use fans. With fans, how do you find the component temperature? From this equation:

$$T_{brick} = T_{inlet\ air} + Power/(Geometry\ and\ Air\ Flow) \qquad Eq.\ (6\text{-}4)$$

"The brick temperature depends on four things. It starts with the inlet air. Obviously, the higher the inlet air, the higher the brick temperature. Then there is the power given off as heat by the brick. The more power dissipated, the higher the brick temperature. This is fixed, because the U-238 draws a constant amount of power when it is running. **Geometry** is some constant value based on the geometry of the circuit boards, such as the surface area, location of the brick on the board, air flow obstructions and so on. It doesn't change either, unless you redesign the circuit boards. Then there is the air flow. The faster the flow, the lower the brick temperature. But only up to a point. Even if the flow gets infinitely high, like in the automated car wash where the hot air dryer nearly blows the paint off your car, the last term of the equation gets smaller and smaller until it is almost zero. But it can't become negative, so the lowest brick temperature you can get is the same as inlet air temperature.

"The trick is to figure out the geometry factor, which I did by testing. I measured T_{brick}, $T_{inlet\ air}$, air flow and power. Air flow was a little tricky, because the fan vendor understated the flow curve in their specs. But after I bribed the application engineer with a piece of gum actually chewed and spit out by Michael Jordan, he gave me the real pressure versus flow rate curve needed to calibrate my test fixture. Using the data from my tests, I calculated a value for geometry and used it to draw up a graph of allowable flow rate and inlet air temperature that looked like (Figure 6-4):

"The line on the chart tells you the combination of air flow and inlet air temperature that gives you a brick temperature equal to 85°C. Any combination above that line (in the shaded area) is safe, because the brick temperature will be lower than 85°C. Anything below the line will cause the brick to be too hot."

Herbie said, "So why isn't this graph in the spec we give to the customer? Even I can understand this."

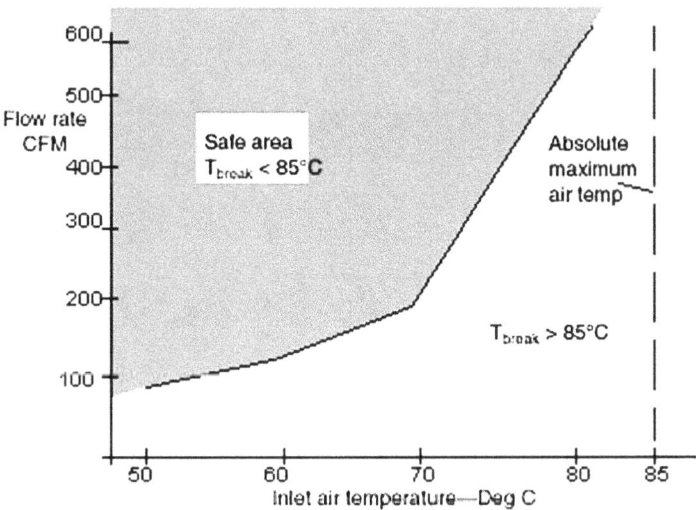

Figure 6-4

"A compromise," I said. "I gave this graph to Armand in marketing and he just laughed. He said customers demand a single number spec, like a maximum air temperature. We argued and in the end compromised. Armand let me have two numbers—a maximum air temperature and a minimum air flow. So we picked one value off the graph, 50°C and 80 CFM, added a 25% safety factor and so you get the spec of 50°C and 100 CFM."

Leon slid the place mat toward him. "Can I keep this?" he asked.

"You're buying," I said. "Seems to me when you're footing the bill, you deserve to get the whole picture.

THINKING KINKS JINX SINKS

"How come heat sinks never do what you expect?" Herbie complained.

"Depends on what you expect," I said. "My shampoo claims to make my hair look fuller, but look what it has to work with."

"Here's what I expect. *You* told me a heat sink works by increasing the surface area of a component. I measured the psychic/optical transformer (POX) chip at 100°C. Then I added a heat sink that had twice the surface area as the chip package. So I expected the temperature to get cut in half, right?" he said.

"Not exactly," I hedged.

"Not exactly is right," he went on. "The POX temperature only went down about 5°C. I was…"

"Disappointed?"

"Apoplectic!" Herbie fussed. "Doesn't anything work the way it's supposed to anymore? Aluminum doesn't have the zip it used to in the good ole days."

"Nah," I said. "Today's heat sinks are better than ever! Electronics these days just put out a lower quality heat—unreliable and hard to predict, you might say a slacker kind of heat. But seriously, there are lots of reasons why heat sinks don't do what you expect."

We spent the rest of the afternoon, skipping even our 2:30 coffee break, to list them.

It only helps the temperature *rise*. Adding surface area to a component by gluing on a heat sink affects only its *temperature rise* above the air, not the whole temperature reading. Herbie expected his 100°C POX to go down to 50°C when he doubled its surface area. But the room air temperature was 25°C, so the temperature rise above air was 75°C. At best, by doubling the area, he could cut that rise in half to 38°C, giving him a POX temperature of 63°C. That explains some of the problem. But wait, there's more!

Too much of a good thing. There is such a thing as too many fins on a heat sink. Once you have decided on the available volume for a heat sink, there is only one way to add surface area—by increasing the number of fins. The more surface area (fins) you add, the more you reduce temperature, up to a point. After that, more fins start plugging up the air passages and air starts to flow around the heat sink instead of through it. Temperature starts to go back up. For any combination of air speed, power and sink volume, there is an optimum number of fins. If you try to reduce the temperature further by adding fins, you could make things worse. At that point, doubling the area may actually make the temperature go up, not down (see Figure 6-5).

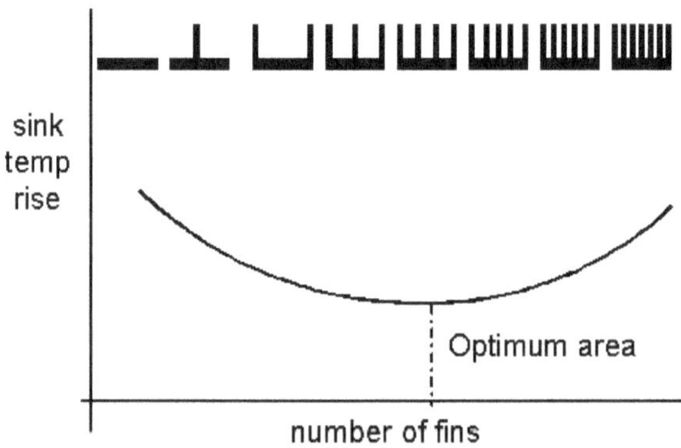

Figure 6-5 Too many fins can be as bad as too few fins

For natural convection heat sinks around 1 watt and 1 inch on a side, the optimum fin spacing is about 1/4 inch (6 mm) (the Pinky Rule—if you can stick the tip of your pinky between the fins, the heat sink will work in natural convection).

Poor attachment. The heat sink has to be in good thermal contact with the component to do its job. Just pressing a bare heat sink to the top of the case may leave a whole layer of microscopic air bubbles between them, and air is a terrible conductor of heat. A thermal interface material, like glue, tape, compressible rubber pad or the notoriously messy thermal grease, can improve this tremendously, if you do it right. I have pulled off poorly performing sinks that were attached with tape, and seen from the pattern of adhesive left behind that less than 10% of the surfaces were actually touching. That means 90% of the joint was a thin layer of air! Tape was not a good choice for that particular component, probably because the top surface wasn't flat enough for the thin tape to fill in all the gaps.

No air flow to start with. Your component may be hot because it is in a dead end of your chassis—with very little air flow. Adding a heat sink won't help much if the air isn't moving. Poke a vent hole instead.

Local air temperature is higher than you think. This is an extension of the first item about temperature rise. The heat sink can only improve on the temperature difference between the component and air—that is, the air the component sees. Herbie's POX chip does not have much of a chance if the air approaching it is already 90°C because it has passed over a bunch of other hot components.

Vendor data sheets may be "optimistic." Sometimes a heat sink catalog will give a single value of $R_{s\text{-}a}$, or thermal resistance between the sink and air. In the fine print the catalog admits that the actual thermal resistance of any sink depends very strongly on how you use it. They test them in a wind tunnel. Will your board be used in a wind tunnel? Then don't trust their $R_{s\text{-}a}$. If you were a vendor, would you report the most optimistic or pessimistic values?

Improving a road that has no traffic on it. This is the most common problem, and the hardest to understand. It is also the hard-

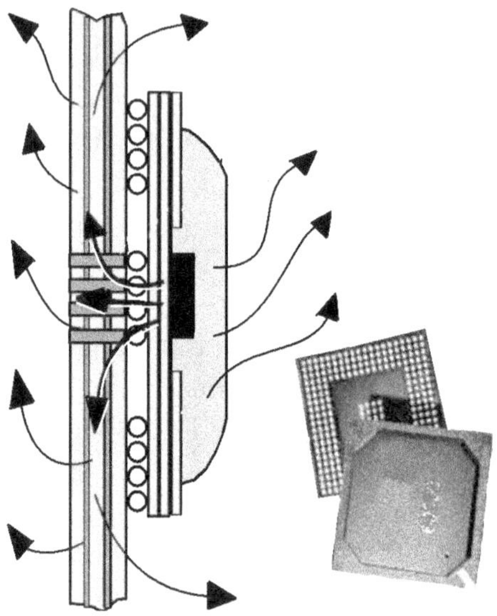

Figure 6-6 Heat takes more than one path out of a BGA package.

est to quantify or predict, because it depends not only on the component you want to cool and the heat sink you select, but on the board
construction and on every other component on the board.

People tend to think that all the heat passes out of a component
directly into the air. In reality, a big portion of the heat flows through
the leads into the board, and then from both sides of the board into
the air (see Figure 6-6).

We can simplify all the heat paths into just two main ones, and draw
them as a resistance network so simple even I can understand it.

Some of the component heat flows through the thermal resistance
of the board into the air, and some through the resistance between
the component case and the air. How it splits up depends on the ratio
of the two braches through which the heat flows. More heat travels
down the path with the lower resistance (see Figure 6-7).

What happens to Herbie's POX chip if the board resistance and
the case-to-air resistance are about equal? Then about half the heat
flows into the board, and half out the top of the case. What happens

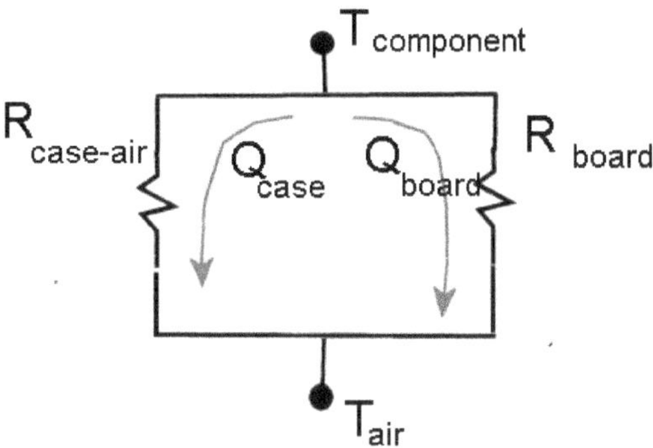

Figure 6-7

when Herbie adds his heat sink, doubling the area of the top of the component?

Doubling the area cuts the resistance between the case and the air in half. But it doesn't do anything to change the board resistance. The POX temperature depends on the total resistance, not just the resistance from case to air. How has adding the heat sink changed the total resistance?

Remember how to add resistances in parallel?

$$\mathbf{R_{total} = 1/(1/R_{board} + 1/R_{case\text{-}air})} \qquad \text{Eq. (6-5)}$$

Reducing $R_{case\text{-}air}$ by 50% only reduces R_{total} by 33%.

It gets worse if R_{board} is much lower than $R_{case\text{-}air}$. Let's assume R_{board} is about 1/10 the value of $R_{case\text{-}air}$. Then you add a heat sink to the case and cut $R_{case\text{-}air}$ in half. R_{total} only goes down by 8%, which means the temperature rise goes down only 8%. You have improved a path that didn't have much heat flow in the first place.

This isn't so hard to understand. But it is nearly impossible to pre dict. Because nobody knows the value of R_{board}. It is not a property of the component, and not a property of the board. It is a combina- tion of all that, the air flow and the other components on the board. Put the POX chip on the same board in five different places, and R_{board} could be very different for each one, depending on the neighboring components. It can't be measured. The only way to figure it out is to use a thermal simulation program like the Therminator, which can calculate the air flow and heat conduction in the board all at the same time.

That's why it's so difficult to make a back-of-the envelope calculation to predict how well a heat sink will work. Even if you do everything right, and, for once, Herbie did do everything right, you don't know how much of a component's heat is flowing into the board, so you can't tell how much its temperature will go down when you add a heat sink.

Here is the translation for the *Variety*-style headline for this chapter, *"Thinking Kinks Jinx Sinks"*: Unrealistic expectations of heat sink performance can lead to disappointment.

THE MAGIC PIPE

6.4

Once upon a time, Hodgepodge the Hedgehog visited the Three Bears. Mama Bear was grumpy. In fact, the whole Bear family was upset in the wake of the recent Goldilocks home-invasion incident. Mama Bear was making breakfast (see Figure 6-8).

"My porridge is too hot," said Papa Bear.

"And my porridge is too cold," said Mama Bear.

Baby quietly gobbled down his porridge. Hodgepodge stared at his bowl.

"We can't go for a walk again," said Papa Bear, "Remember what happened last time!"

"Yes," sighed Mama Bear, "and William the Insurance Weasel hasn't paid off on our claim yet."

"Does this happen often?" Hodgepodge asked.

"Every morning!" chirped Baby Bear.

"Yes," Papa Bear said, "leading to frequent waste of two-thirds of the household porridge budget." He looked at Mama Bear. She covered her face with her apron and sobbed.

"It's not my fault," she said. "I follow the traditional Three Bears recipe for porridge, but no matter what I do, Papa's is always much

Figure 6-8

hotter than mine, and they both stay that way for a long time, given the high thermal capacity of thickly cooked porridge and heavy crockery bowls."

"Don't fret," Hodgepodge said. "My cousin Paulie the Porcupine is a kindly Wizard. He showed me how to make a Magic Pipe that just might help."

"No smoking your funny pipe in front of the kid," Papa Bear growled.

"Not that kind!" Hodgepodge said with a grin. "A Heat Pipe! A bit of copper, a splash of water, a few magic words and the heat will practically jump out of Papa's porridge into Mama's porridge."

Papa Bear scoffed and went back to reading his *Medieval Times.* Mama Bear began scrubbing the old porridge pot. Only Baby Bear followed Hodgepodge out to the village blacksmith forge, under a spreading chestnut tree.

Using only the 15th century technology available in most fairy tales, Hodgepodge fashioned a tube out of copper sheet. On the inside of the tube he scratched thin grooves from one end of the tube to the other. Then he brazed one end of the tube closed.

"Now for the magic substance that makes it work!" Hodgepodge said with a wink. He dipped a reed in the quenching bucket and let a few crystal clear drops fall into the open end of the tube.

"That's just water!" squeaked Baby Bear.

Hodgepodge smiled. "Isn't water magical? One day it's a lake and the next it's a cloud," he said. "Now I have to suck out most of the air from the tube while I'm brazing it shut. The tricky part is not burning your lips!"

Using Paulie's proprietary process, Hodgepodge sealed the pipe, leaving inside a partial vacuum plus the few drops of liquid water.

"What's so magic about this?" said Baby Bear, holding it in his paws. "Where do you put in the batteries?"

"No batteries," Hodgepodge answered. "It doesn't need any kind of power at all. The heat from the hot porridge makes it go."

When they got back to the Three Bears' house, the breakfast porridge was on the compost heap, and Mama Bear was serving the lunch porridge.

"It's Magic Time!" Hodgepodge announced. "Papa Bear, your porridge?"

"Too hot."

"And Mama, your porridge?"

"Too cold."

Hodgepodge placed the bowls near each other, bent the copper tube into a U-shape, then shoved one end into each batch of porridge. Within a few minutes, the hot bowl had cooled down and the cool bowl had warmed up, and for the first time since the H. C. Anderson Memorial Spaghetti Dinner, Papa and Mama enjoyed a meal together (see Figure 6-9).

"It's magic!" said Baby Bear.

"No, no," Papa Bear said. "Copper is a good conductor of heat. That's how this clever feat is done."

"The Magic Heat Pipe conducts heat many times better than a hollow copper tube. Even many times better than a solid copper rod of the same size," Hodgepodge explained. "This tube has three special tricks that make it work like magic:

Figure 6-9

- A little bit of water
- A partial vacuum
- Grooves inside the tube

"Because the pressure is low inside the tube, the boiling point of the water is very low. When I put the end of the Heat Pipe in the hot porridge, the liquid water turns to steam. The steam flows through the tube, looking for a place where there is less steam, such as the cold end, inside the cold porridge. There the steam turns back to droplets of water. They collect in the grooves in the walls and slide back along the walls by capillary action, looking for a place where there is less liquid water, such as the hot end. The water droplets turn to steam again, and the cycle goes on and on until there is no more temperature difference between the two ends of the tube! The phase change of the water can carry much more heat at a lower temperature difference than copper can move heat by conduction."

"Phase change?" said Mama Bear.

"Capillary action?" said Papa Bear.

"You're right, Hodgepodge," said Baby Bear. "Let's just say it's magic."

Mama Bear sniffed the heat pipe, suspiciously. "Can this magic pipe be used in other ways, say, to equalize the temperature of bowls of soup, or even compote?"

Hodgepodge rubbed his furry chin. "I suppose it could be used to move heat from any hot spot to any cold spot. Paulie says it even works against the direction of gravity. He once spun a fable about how in another kingdom he used a magic pipe to connect a hot microprocessor chip to a heat spreader plate underneath the keyboard of a laptop computer."

"Hah, wizards!" Papa Bear said. "Why couldn't he just attach the micro-whatever directly to the heat spreader plate?"

"Why didn't you and Mama Bear just mix your porridge together in one big bowl?" Hodgepodge said. "The simplest thermal solution isn't always the best way to enjoy the whole meal. The heat pipe gave the designers the freedom to place the processor exactly where it ought to be for electrical and manufacturing reasons, and still allow the heat to get out to the fingers of the laptop user, where it belongs."

Papa Bear nodded and patted his full tummy. Mama Bear proclaimed, "Excellent magic! "Hodgepodge, you must stay with us for a whole week. By the way, how are you at fixing furniture?"

WHEN 6% IS 44%

I love power bricks.

Well, not love exactly. Of all components I hate power bricks the least (see Figure 6-10).

Here are some reasons why, from the point of view of a thermal engineer:

- I can understand what a power brick does: 48 volts goes in and 3.3 volts comes out. I can see the point of that on a circuit board. At least it's more obvious than a component called clock and data recovery (CDR). Seems to me that if they put the clock and data in the same place every night when they get home from work, like on the kitchen counter by the fridge, they wouldn't keep losing them and need all those CDRs to recover them.
- I can figure out the power dissipation of a power brick all by myself. You measure the current and voltage in, and the current and voltage out. Current × voltage = power. The difference between the power in and the power out is the heat generated

Figure 6-10

in the brick. Or you can get it from the input power and the efficiency rating from the component spec: Efficiency = power out/power in. Easy.

- The temperature limit for a typical brick actually means something. It usually says, "Operating limit: 100°C, measured on the base plate." This is something I can measure while the brick is powering a board, and I can tell right away if it is OK. Compare this to a typical operating temperature limit for an SDRAM chip: "Max temperature 70°C ambient." Where do I measure "ambient"? Nobody can tell me.

- The last reason is an aesthetic one, almost moral or spiritual in nature. It might be compared to this feeling: You spend hours preparing a gourmet dish—and you are proud that for once you combined the subtle flavors perfectly—and you serve it your family, and they proceed to douse it with ketchup and salt, then scarf it down while watching *The Simpsons* on TV. Mechanical engineers are ingrained with the notion that high efficiency is good and waste is bad. We are taught to design machines that reduce losses to the bare minimum. (Even now I hear Professor Chu intoning, "*Mister* Kordyban, do you think enthalpy grows on trees!") It is our sworn duty, from the Archemedic Oath, to give our customers the maximum output power for the minimum input power. When I worked as a co-op student for the Ford Engine research department, I learned that a typical automobile engine could convert only about 25% of the energy in gasoline into useful power at the drive wheels of a car. The rest went into heating up the atmosphere. I thought that was horrible, until I entered the world of electronics. Here a microprocessor consumes 100 watts of useful electricity, and turns about 99.99% of it into heat. OK, it may be doing something desirable, like calculating a spreadsheet for your budget, or updating the display of your latest shoot-'em-up game. But when it comes to output power divided by input power, its efficiency is as close to zero as makes no difference. To put it simply, power bricks appeal to my sense that efficiency is good.

A good power brick (or DC/DC converter, as they are sometimes called) can be almost 95% efficient, a poor one as low as 75%. But even that is pretty good compared to a laser transmitter. A laser can draw 1 watt of electrical power and spit out a meager 10 milliwatts of laser light. That's about 1% efficiency. You're not going to blast through a hull of solid neutronium with that kind of delivery.

There must be a law of human nature, then, that forces us to take for granted the ones we love. That can be the only reason why I was recently fooled by the thermal behavior of a power brick.

Jim had left a yellow sticky note on my coffee mug. He used to stick them on my computer screen, but found that he got a quicker response with the mug. "I found another brick," it said simply.

That meant I was supposed to come to his lab and do yet another thermal test on a sample power brick. It had started almost a year before, when I did a thermal test on his board, the pigeon substitute cortex (PSC). He was attempting a major cost reduction of the human brain unit for use in Internet data switching. Turns out that the average intelligence level in Internet data can be handled by a brain much smaller than a human one, so Jim was using one from a pigeon. Anyhow, the only thermal problem I found was the power brick on the PSC. It was 98°C at worst-case conditions.

"The limit is 100°C," I told him, "so technically, you are OK. But your margin is very small. If your power needs to go up even a little in the future, your brick could go over its operating limit. So I'm not flunking your PSC, but I wish you could do something to get the temperature down in the future."

Jim said nothing, nodding, and went back to his pigeons. But week after week, he found another brick vendor, got another sample and asked me to test it. Week after week, I came back with the same results, "2 to 4°C under the limit."

So it was not with the utmost eagerness that I toted my thermocouples to his lab that day. "What makes this new brick so special that it is actually worth testing?" I asked, shoving aside a pile of used samples to make room for my meter.

Jim silently held up a data sheet. He had circled the efficiency number with a red marker: "90%. Big deal," I said absently. "Your first brick was 84%. I don't think an improvement of 6% is going to make a dramatic difference in the brick temperature. When I was in high school, that was the difference between a B-minus and a B."

Jim shrugged and offered me a static bag containing the PSC board. I sighed and went to work attaching my temperature probes.

An hour later my lack of enthusiasm had changed to frustration. "Jim, that PSC you gave me must be broken. The power brick is much too cold!" I said.

Puzzled at my exasperation, he ran a diagnostic. All the lights came up green. He measured the input power for me. It was low, but not too low. The pigeon was cooing at the proper frequency. As a last check I put my hand on the board. The neural transceivers were all appropriately hot to the touch.

"But this can't be right!" I said. "The brick is only 77°C. That dinky change in efficiency could not have made it come down over 20°!"

Jim cocked his head to one side, as if to say, "Or could it?" That made me do some calculations on paper instead of in my head.

First of all, I needed to be thinking about the temperature rise above ambient, not just the temperature of the brick. I did all the testing at the same ambient, 50°C, because that was the maximum required for the system. Then I did a little ciphering to figure out just how much the heat generated inside the brick had changed (see Table 6-1).

Table 6-1

Brick	Efficiency	Input (watts)	Output (watts)	Heat (watts)	% change
Old	84%	29.5	24.8	4.7	—
New	90%	27.5	24.8	2.7	–43%

Table 6-2

Brick	T_{case}	ΔT above ambient	% change
Old	98°C	48°C	—
New	77°C	27°C	–44%

So even though the efficiency increased only 6%, the heat generated went down by 43%. How did that compare to the temperature rise? (see Table 6-2)

Everything fell into place. After looking at the two tables, it all seemed so obvious. A small change in overall efficiency made a big difference to the amount of heat generated, because the amount of heat generated was relatively small to begin with.

I showed the tables to Jim. "You kept pecking away and finally you found a brick that works. In this case a 6% improvement gives us a tenfold increase in margin under the operating temperature limit," I said.

This little numbers game was a valuable lesson for me. My professors were right—efficiency is good. And in some cases, even a small increase in efficiency can have a payoff beyond expectation.

That was reinforced when I got back to my desk and found another sticky note on my coffee mug. This one simply said, "Thanks." Plus, the mug had been filled with fresh coffee.

SO CRAZY, IT JUST MIGHT WORK!

Heat sinks and fans. Heat sinks and fans. Electronics doubles its processing speed every 18 months and shrinks in size by at least half every year. It seems like the electronics people are zooming along with technical innovations, and us thermal and mechanical people are standing around with our thumbs stuck in our thermal epoxy. Heat sinks and fans are all we have to offer. OK, CFD was introduced in the 1990s. That got us out of the "wet finger and slide rule" days. But isn't anybody working on some new technology for cooling electronics? After all, if they keep doubling the speed and cutting the size of the components, pretty soon heat sinks and fans won't be able to do the job.

University and industrial labs are working on some novel technology for cooling electronics. Here are some of their new ideas, which I gleaned from attending an international conference on electronics cooling. Is it a peek into the future of the thermal business? Most of these developing concepts will never make it into a real product. Which ones will become useful in the future, only time will tell. But this will give you an idea why we tend to stick with heat sinks and fans right now.

Bold, Innovative ... But Not Quite There

The following ideas are so new and creative that nobody is using them yet in real products. Or maybe there is some other reason.

Low-Melting-Point Alloy Interface Materials

One roadblock to using higher and higher power components is the thermal resistance between the top of the component package and the bottom of the heat sink. Phase change materials (melting wax) and thermal grease are improvements over thermal glue, but what if you had a thin metal foil that could melt at a relatively low temperature so it could flow into the microscopic cavities in the surfaces of the package and the heat sink, virtually welding them together? This dream has been realized, at least in the lab. Researchers cooked up a metal alloy that melts a few degrees below the maximum operating temperature of a microprocessor. They formed it into a foil and clamped it between a heat sink and processor package. After heat from the processor melted the foil, the interface resistance became very low. For a while. Under temperature cycling (which happens by turning the computer on and off each day), the package and heat sink expand and contract, squeezing the interface material and squirting it out of the joint all over the circuit board. Not only did the thermal resistance go back up, but the rest of the board did not respond well to random splashes of electrically conductive metal.

Acoustic Refrigeration

From the title of this presentation, I was hoping for a gizmo that could convert thermal energy into acoustic energy, so that component heat could just float away to the tune of *"Brahms's Lullaby."* What it turned out to be was a conventional vapor compression refrigerator small enough to fit on a circuit board. By small, I mean that it takes up 20 to 30% of the space on a large motherboard. The conventional rotary compressor is replaced by an electromechanical resonator (a speaker) that compresses the refrigerant by creating a standing wave in a resonant chamber. Its main advantages are that it is small enough to mount on a board, and it can handle a variable load. The acoustic refrigerator was able to control the temperature of four microprocessors on a single board, even though their total power could vary over time from 10 to 400 watts. At the end of his talk, the presenter admitted that the acoustic refrigerator was not yet in production, because of a few nuisances like the "loud" noise it produces, and reliability issues due to vibration-induced fatigue.

Heat-Driven Cooling Fans

At the end of a long paper on optimizing heat sinks for notebook computers, the presenter hinted at an idea for saving energy in the cooling system. Saving energy is always important when the product runs on a battery. A notebook microprocessor (at the time of his study) typically puts out about 25 watts of heat, and, to keep it cool, needs a fan that eats about 0.3 watts of electricity. Why not use the waste heat from the processor to create the electric power for the fan, using a thermo-electric (TE) generator? A TE generator is the same thing as a thermo-electric cooler, just run backward. It is a solid state device that converts temperature difference into electricity, or electricity into a temperature difference. Even with a dismal efficiency of 2 or 3%, a TE generator could convert the 25W of processor heat into enough power to run the cooling fan (see Figure 6-11).

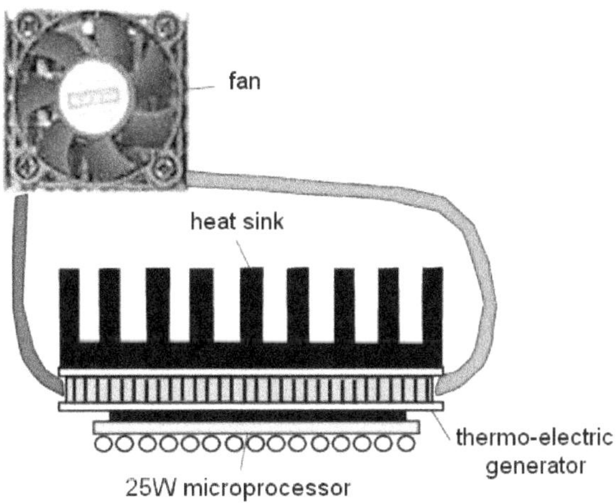

Figure 6-11

The idea is not fishy, but its application might have a catch: The processor runs hotter because of the extra thermal resistance of the TE generator, which could hurt long-term reliability.

Saving 0.3 watts may be significant for extending battery run time for a laptop, but I don't see much use for this idea in non-mobile applications. Unless you take the electricity from the TE generator to run a TE cooler, and then…

Recirculating Air Flow Through the Shelf—on Purpose!

Sometimes when I am analyzing a fan tray, I spot areas where air is flowing backward through the shelf, or going around in circles (see Chapter 2.3). This is generally a bad thing. The air gets hotter and hotter every time it passes heat-generating components, and the hotter the air gets, the hotter the components end up. So I work hard to get rid of recirculation. One presenter, though, has been trying to harness the power of recirculation for good instead of evil.

His telecom shelf is a mid-plane design, with high-power line cards plugging into the front and low-power I/O cards plugging into the back (see Figure 6-12). The overall power of the shelf was not large, but several components on the line card had extreme power that needed high-speed air flow. The fan tray was just not capable of producing the high speed he wanted in the available space, without it howling like a werewolf during the full moon. But if he connected the inlet and outlet of the fan tray in a closed loop, he could get much faster air flow. Of course, that would mean the air would just get hotter and hotter as it went around in a circle. He added leaks, so cool fresh air could mix into the system and hot air could leak out. Using flow network modeling software, he was able to balance the resistances of the inlet and exit vents just right, so that the internal air speed was as high as he wanted, and there was enough leakage through the system to keep the air temperature rise acceptable.

This presentation was a great triumph for flow network modeling software, demonstrating how it can be used to design even a complex cooling system like this. Whether this idea is brilliant or nutty, I haven't figured out. It can definitely work, and maybe it even has advantages. But my gut feeling is that it is a very unstable cooling

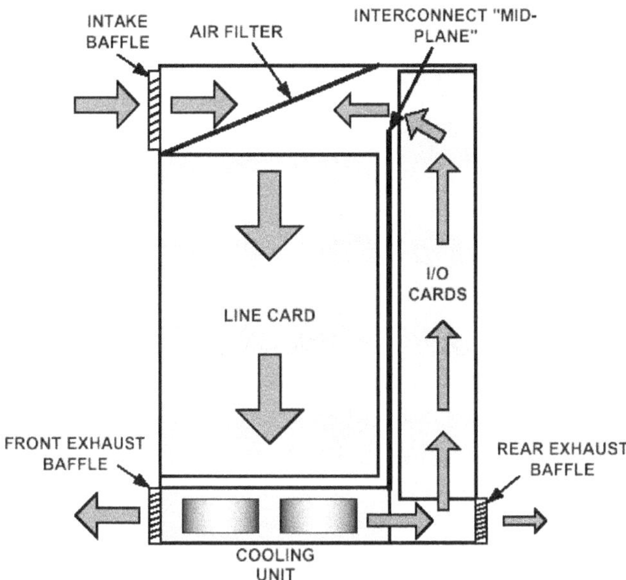

Figure 6-12

system. Balancing the inlet and outlets against the fan performance to get just the right mix of recirculation and leakage is like balancing a pencil on its point. You can find a solution that works, but any small disturbance could knock the whole thing over. What would happen during a fan failure, or if the shelf were only partly populated or if the filter clogged up? How easily could the balance be thrown off and cause thermal runaway? And could I use the ever-increasing air temperature to run a TE generator to increase power to the fans and...?

New Thermal Technology with More Promise

My first reaction to the following presentations was, "Huh?" But soon I was thinking, "Aha!"

<u>The heat sink that melts on purpose</u>

The metal-spitting interface people from the first idea in this chapter came up with another use for their low-melting-point metals.

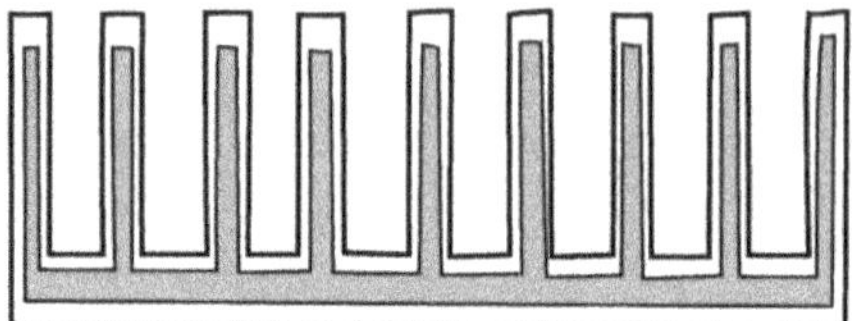

Figure 6-13 A cross-section of a melting-core heat sink. Melting absorbs heat without a temperature rise.

They hollowed out an aluminum heat sink and filled it with their stuff (see Figure 6-13). At this point you're saying, "Huh? So they made a heat sink that melts in the middle. How does that work any better than a simple solid aluminum sink that costs a lot less to make?"

In most applications, in fact, it doesn't work any better. But it has one pretty neat feature that comes in handy in one very specific application. Unlike the solid sink, it can absorb a significant amount of heat without increasing in temperature (at least for a while.) Don't worry, the Law of Conservation of Energy is still in effect. That heat energy still goes somewhere. It goes into the Latent Heat of Fusion of the melting-metal core.

Figure 6-14 shows what happens when you heat up a solid. Its temperature goes up steadily until it reaches the melting point. Then, even though you keep pumping in the heat, the temperature stops changing. The energy goes into changing the solid to a liquid, and not into increasing its temperature. The temperature won't increase until the material is completely melted. The energy absorbed during melting is called the latent heat of fusion.

Latent heat can be pretty large. It is one reason why we fill picnic coolers with ice instead of, say, rocks chilled in the freezer. It would take 100 pounds of –40°C rocks to equal the cooling power of 10 pounds of melting ice.

Before we get to the "Aha!" stage, you need one more thing—the application that this is good for. Suppose you have an electronic component that has very high power dissipation for short periods of time, but most of the time the power is relatively low. One common example of this is a motor control device for a robot arm. It might have to

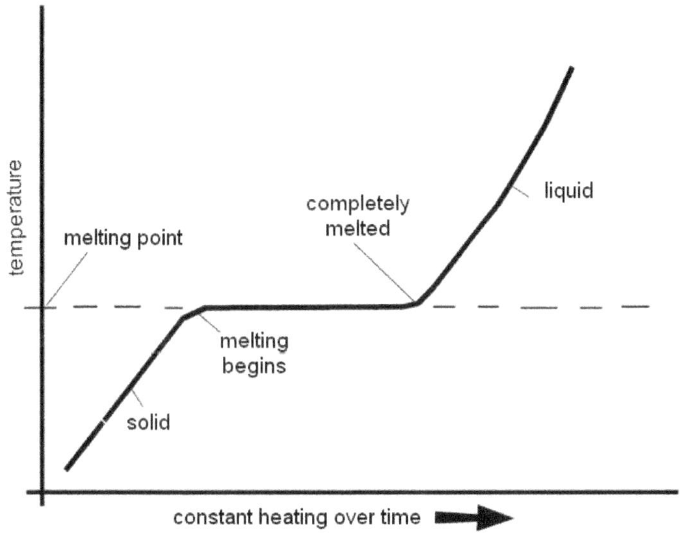

Figure 6-14

deliver 100 amps for 5 milliseconds, and then nothing for another 5 minutes. The duty cycle is very spiky, not continuous. Another example is a workstation microprocessor. Most of the time it is processing mouse clicks, dissipating 20 or 30 watts, and then suddenly it is doing an enormous set of floating point calculations, dissipating 150 watts for a few seconds. In each case the average power is fairly low and a simple, small heat sink could handle it. But the peak power overwhelms the small sink and the chip's temperature spikes over its operating limit for a short time. Sudden large changes in temperature can be very damaging.

A massive solid heat sink could help smooth out these temperature spikes. But the beauty of the melting-core heat sink is that flat spot in the temperature curve. Let's say your component has a maximum operating temperature of 90°C. You select a core material for the sink with a melting point of 85°C. You can keep hitting that component with power spikes, and until the core is completely melted, the sink temperature can't go above 85°C. And if the sink stays at 85°C, the component can stay below 90°C.

You run into a problem only if the core melts completely and you keep pumping in heat. The temperature of the heat sink and the component will start going up again, and this sink is no better than a solid one.

For a component with high momentary power, a melting-core heat sink can be much smaller than a solid one. The core absorbs the peak heat load by melting, and then the fins release that heat to the air during the longer periods of low power, allowing the core to refreeze. A solid sink would have to be oversized to handle the peak power.

The melting core acts like the retention ponds you see around suburban office campuses. The sewer system is sized to handle only an average rainfall. During a cloudburst, water backs up into the retention ponds. After the storm, water is released into the sewers at a rate they can handle. That way we don't need to have 20-foot-diameter sewer pipes under every street to handle the peak flow.

Now that you know about melting-core heat sinks, you can start thinking about how component power varies with time. If you have a high power component that is spiky in a fairly predictable way, maybe you can use a melting-core heat sink. The spikes have to be somewhat predictable, so that the core can be sized to make sure it never completely melts. For example, if your SDRAM has a 10% duty cycle (active 10% of the time, idle 90%), that may be a good place for the melting core. If the duty cycle is 10% today, but maybe 90% tomorrow, depending on the usage by the customer, that is probably not a good place.

The Price of a Watt

Another presentation recommended that large telecom offices and data centers could save power by generating electricity on site and using the waste heat to run absorption-cycle air-conditioning. That isn't something that electronics cooling engineers think about, but the presenter also gave a big-picture view of just how much energy goes into running electronics (see Table 6-3). The next time you add a 1-watt component to your board, think about where that watt will come from (see Figure 6-15):

Table 6-3

Power Consumer	Efficiency	Power
"Useful" component power	—	1 W
Air-conditioning to remove 1W	COP = 7	0.14 W
48V to 3.3V conversion	85%	0.18 W
AC to 48VDC conversion	85%	0.21 W
AC transmission lines	80%	0.38 W
Fuel-to-electricity conversion	30%	4.46 W
Total fuel required	—	6.37 W

Data centers are even more wasteful than telecom offices because they bring in AC power, convert it to DC to charge batteries, then convert battery output back to AC to distribute to the servers, which then change it back to DC again. Telecom equipment runs directly on DC power, so one of the lossy conversion steps is skipped.

That itsy-bitsy watt you add to your design ends up dumping more that 6 watts of heat into the atmosphere someplace, not to mention the CO_2, the coal dust and/or the plutonium involved. And even if none of that bothers you, the total fuel in the end determines the electric bill our customers have to pay to operate our products. Every watt left out of the electronics saves 6 watts for the world, and makes my job easier, in case that matters to anybody.

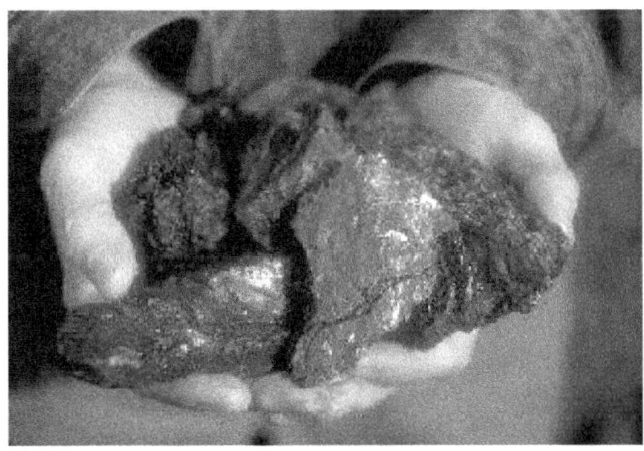

Figure 6-15 A 1-watt part and the fuel it takes to run it.

Section 7

Telecom: A Field With Myths and Mistakes All Its Own

The telecommunications industry is a world all its own. For example, (in the United States, at least) the telecom standard color code for DC power wires says negative is red and positive is black, when the rest of the world uses red for positive and black for negative. This can be quite shocking to an engineer coming from the consumer or automotive electronics industry.

Cooling electronics in the telecom world has certain unique quirks as well. The thermal design challenges are not so much driven by extremes of power dissipation, crazy environments or even cost, as they are overshadowed by the need for the phone network to never, ever stop working. Reliability, availability, repairability and redundancy drive everything. If you need one fan to cool your circuit, then you probably need two. But having two creates potential leaks if one dies, so then you probably need four. And just in case, better make it six, or even eight.

The companies that provide telephone service are used to buying a system, installing it in the network and letting it sit there, switching traffic and generating revenue for the next 40 years. It is supposed to do that during earthquakes, power failures, air-conditioning faults, unattended, without ever being turned off. Now you can begin to understand why the telecom industry balked so long at introducing fans at all into their cooling schemes. There is no such thing as a fan that runs steadily for 40 years.

This is fair warning that the following chapters are even more heavily laced with telecom standards and telecom acronyms and telecom design concepts than the previous sections. Here is your chance to skip to the end without having to read about NEBS about 20 more times. But for those of you who slave away in the telecom biz, and for those of you who only dream of slaving away in the telecom biz someday, here is the section that will make you feel the most at home. Here are your nine nines of reliability, your 50°C ambient, your Zone 4 earthquake test. For the rest, there are still some general heat transfer ideas worth learning, a few more details about the human brain unit and a couple of funny pictures.

Thinking Inside the Box

Herbie came back from the trade show with a shopping bag full of promotional junk. He had the dodecahedron desk calendar, the satellite phone Slinky and, my favorite, a pair of tin cans linked with a fiber optic cable. He dumped the whole bag out on the floor of my office and pawed through the pile until he found a particular brochure (see Figure 7-1).

"Here's something you'd be interested in. It's a great new concept for fan cooling that saves a *ton* of space in the rack. Plus it will let us fit in a lot more boards, which is something we're always looking for," he said, spreading out the color brochure on my desk.

"What the heck is this thing?" I asked, turning the paper sideways, and then upside down.

"It's the latest router from Rimrock Router Systems," he explained. "It's getting harder and harder to tell the difference between phone service and Internet traffic, with the router folks getting into telephony and the telephone folks sticking their noses into data."

"And this thing mounts in a telecom rack?" I asked.

"That's what they say," Herbie said. "One accessory kit is a set of mounting brackets for a telecom-style rack. And the tiny type on the

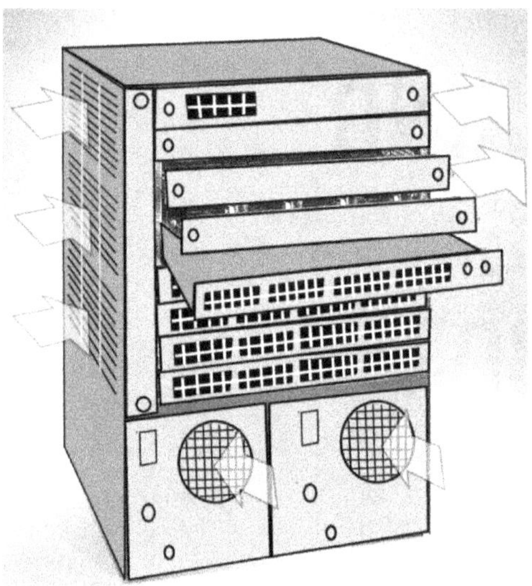

Figure 7-1 In this example of data product design, even the different design departments couldn't agree on a single air flow direction. The router boards are cooled left to right, and the power supplies on the bottom are cooled front to back.

back of the brochure says: *'Telecom ready! Designed to comply with a full 58% of NEBS[1] requirements.'*"

"But where do you put the brackets?" I asked, "This picture looks like the box fell over on its side."

"It is right side up in the picture. That's the beauty of this thing," he said.

"Beauty?" I said.

"The boards are mounted horizontally! How's that for thinking outside the box?" he answered. "The fans on the left side suck air in, blow it through the boards and then out through the vents on the right side. The beautiful part is that it doesn't need any of those gigantic, space-grabbing baffles between the shelves. You can stack these boxes up in the rack edge-to-edge with absolutely no rack units wasted on sheet metal that does nothing but steer hot air around. This is really great if you have customers who may want only one or two boards. Then you can make the box very short, and take up a minimum amount of space in the rack. If you put the boards vertically, like we normally do, the shelf has to be as tall as one board. If you equip only one or two boards, the rest of the shelf is empty, wasted space."

I picked up the brochure and filed it in an already overflowing folder. "You were right that I'd find this interesting. I'm going to add it to my collection of examples of ways *not* to design a cooling system for telecom application," I announced.

Herbie rolled his eyes and said, "You old-timers from the Ma Bell system just don't like change, not even when something superior comes along."

"I'll compare birthdays with you anytime, Mr. Vacuum Tube Hobbyist," I said. "I'll admit the old Bell System was stodgy and slow to change. But back when the Bell System *was* a system, they had people whose only job was to figure out how the pieces of the network puzzle had to fit together. They came up with a standard way of putting boards in shelves, shelves in racks, racks in aisles and aisles in rooms in such a way that they all fit together with the room ventilation and lighting and cabling systems.

"There's a reason why we traditionally mount our boards vertically. There is a preferred direction for air to flow through each shelf— either from bottom to top, or from front to back, or some combination. We don't try to flow air side-to-side, because the racks are going to be put together in long lineups side-by-side. If I want to suck in air on the left, there needs to be some empty space there. Most likely there will be a rack upright, or the solid wall of a shelf in the neighboring rack. Worse yet, if there is another shelf blowing air from side-to-side, I might be sucking in its hot exhaust air! The exhaust of my shelf will be blowing against the solid wall of a shelf in the neighboring rack, too, or maybe into the intake of other equipment. This is obviously not good.

"When the army makes camp in a new spot, the first thing they do is dig a latrine. Why? Because they don't want the soldiers, uh, discharging their waste, so to speak, everywhere that they'll be sleeping, eating and doing push-ups. They pick one spot to dump so they don't contaminate everything else. That's what the system engineers for the telco central office have done, only they are dealing with waste heat instead. They designed the lineups of racks so that the fronts face each other in the so-called *front,* or *equipment aisles,* and the backs face each other in the *wiring aisles.* These aisles alternate, as shown in Figure 7-2.

"The ventilation system is built to pour cold air into the front aisles. The shelves are supposed to draw in the cold air in the front and dump hot air into the wiring aisle or out of the top of the rack."

Herbie looked as if I had just stuck a pin in his circus balloon. He said, "But for my one little shelf, this old-fashioned method is such a waste of space! It's so inefficient!"

I nodded and said, "You're probably right. Nobody claims that the army is very efficient at using each person to his or her full potential. But when it comes to making thousands of soldiers do something all together at a command, they are very good at that. And that is what you are trying to do in a central office—make hundreds or thousands of circuit boards work together in a controlled, predictable way.

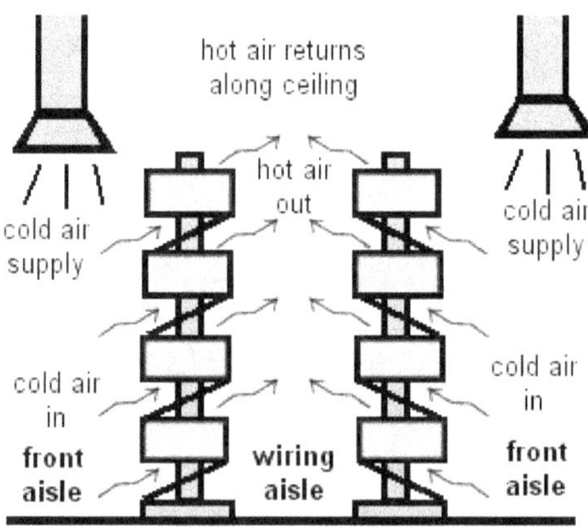

Figure 7-2 The typical central office uses the wiring aisles for getting rid of waste heat.

"On the other hand, you've got your data product vendors like Rimrock. They are less like an army and more like a horde of family campers descending on a state park on a holiday weekend. Each family sets up its own camp however it wants. That works OK until one camper builds a fire and the smoke drifts into his neighbor's tent. Another dumps out wash water that splashes onto somebody's sleeping bags. I park my giant recreational vehicle in just the spot that blocks your view of the lake. Nobody takes responsibility for anything that happens outside the boundary of their little camping spot. What you end up with is a bunch of unhappy campers.

"You talked about thinking outside the box—the router designers are the ones who stop thinking right at the edge of their own box. They put the air vents on the left side, right side, front, back, wherever they need them. Then it's up to the customer to figure out how to make them work if they want to put more than one of them in the same room."

Herbie chuckled and said, "Can you imagine their customer service calls? *'Hello. Your server stopped working because our router is blowing hot air on it? You'll have to contact the server manufacturer for help with that. Oh, they told you to turn the racks around so the server was blowing hot air on our router instead? And the router overheated? OK, the answer is: don't do that. Turn them back the way they were. And thanks for calling. Every call is important to us.'"*

"Nice Lily Tomlin impression," I said, glad Herbie was getting it. "Now what else did you find for me at the trade show that we could learn from?"

"OK," he said, pulling out another brochure, this time with the TeleLeap logo all over it, "If you didn't like the last one, you're really going to hate this one. It's a sideways air-flow box we are private-labeling from another company. It's a telepathy over IP (ToIP) box to connect the HBU to the Internet. Good luck on getting it to work with the rest of our system."

I sighed, "Oh, wonderful! Let me have that. I'm going to the men's room." I stomped off to find a comfortable stall to study the brochure, when I could do a little of my own thinking inside the box.

Notes

NEBS stands for *Network Equipment Building Standards.* It is the industry standard in the United States for indoor telecom equipment. It defines mechanical and environmental requirements for the buildings as well as for the equipment that is supposed to go in them, so they match. The standard is a proprietary document of Telcordia.

"Just Slap It in an ETSI Cabinet and *Voilà!*"

"Welcome to the BRRR division of TeleLeap," Slim greeted me. Some smart guy in the 1990s had established the BRRR division in Iceland as a way of getting a foothold in Europe, to ease selling into the European Union. It wasn't until later we found out that Iceland isn't actually a part of Europe.

The BRRR division specializes in telecom products for transoceanic applications. They started with terminating equipment for undersea cable, and expanded into satellite. They have the patent on the circuit that adds the familiar hiss and echo to the voice lines, so customers feel like they are getting their money's worth when paying for an overseas call.

"Not that I mind visiting Iceland in January," I said, "but if you've had your system in the field for years already, why did you ask me to come and do a thermal test on it now?"

We shed our scarves and boots and parkas and wool sweaters, and even then, Slim was quite portly, despite his nickname. "I'll show you," he said, leading me into the lab.

He pointed to an open rack with four shelves full of circuit boards (see Figure 7-3). "This is the BK1200," he said, "It was designed for

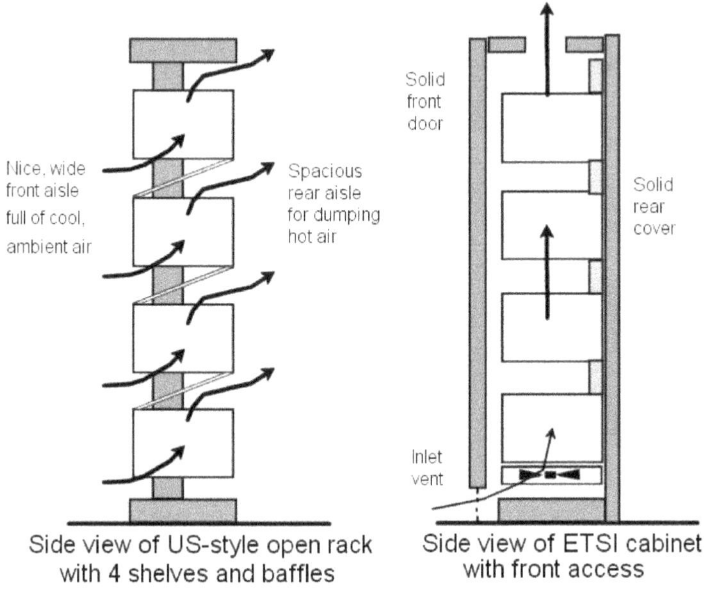

Figure 7-3

use in Canada and the United States, so it is an open rack, 12 inches deep, and it complies with the North American standards from Telcordia. There are no fans, because it works OK up to 50°C ambient without them."

"I see you have baffles between the shelves," I pointed out.

"Oh, *ja,*" Slim nodded, "We found out a long time ago that it was necessary to keep the hot air from the bottom shelves from making the top shelf too hot."

I placed my hand on the cover plate of the top shelf and made a thoughtful "hmmm" sound, to show what an expert I was in temperature sensing. "Feels nice and cool. Those baffles are doing a good job," I said.

"Your praise is quite generous," Slim said, smiling, "considering that this particular rack is not powered up at the moment."

"Uh, right. So what am I supposed to be testing?" I said, shoving my hand quickly into my pocket.

"The BK1200 is very successful all along the North American coast. But now our marketing would like to start selling it on the European side of the Atlantic. Europe needs the exact same hardware, except with a half-twist in all the cables to switch from right-hand to left-hand drive for England. But nowadays, to sell in Europe, it is an advantage for the equipment to meet ETSI specs," Slim said.

"Oh, no, not the European Telecommunications Standards Institute! Those documents are so complex they make Telcordia specs look like the Dick and Jane reader," I moaned.

He showed me the ETSI version of the BK1200 on the other side of the lab. "The main difference between ETSI and the Telcordia version is the cabinet," Slim said, "They want all the cabinets in the office to be accessible only from the front, because there is no aisle in the back. Cabinets mount with their backs against a wall, or even flush back-to-back with other cabinets."

"That's not good," I said.

"Yes, I know," he continued. "First we had to redo all our cables so they connect in the front, because the back of the cabinet is all sealed up solid. Then we realized the baffles wouldn't work, because there was no place in the back to deflect the hot air to anymore."

"Plus, I see that your cabinet has a solid front door, so no fresh air can get in between shelves anyway," I mentioned.

"Marketing insists that customers want it that way," Slim said.

"So the air comes in through this grill at the very bottom of the cabinet," I said. "Then it flows up through all four shelves and then out the top. Didn't you tell me you already knew this would make the top shelf too hot?"

"You bet," he said. "That's why we put this fan tray here at the bottom of the cabinet. We are hoping that these six fans will give us enough air flow that even the boards in the top shelf will still be OK."

"Fans, eh?" I said, rubbing my chin, "I suppose we'll just have to take that 1 in 10,000 chance!"

"Ooooh—that's Captain Kirk in *The Naked Time,* my favorite episode of *Star Trek.* I'll get the coffee started, you start that thing you do with the thermocouples."

Over the next several days I measured dozens of component temperatures in the BK1200 shelf under all kinds of conditions—fans off, fans on, door open and closed, fans at half-speed, fans failed one at a time. It seemed like I was in that lab all night, and the winter nights in Iceland can be 20 hours long.

When I was done measuring, it took another winter's night to process the numbers, extrapolating the component temperatures to the worst-case conditions and comparing them to their operating temperature limits. When I was done monkeying with the spreadsheets, getting the borders and shading just so, I sat down with Slim in his teak- and lava-rock-lined office.

Slim flipped through my 24-page report. "Not good," he said. "All I wanted was a one-page memo saying 'It works,' with your signature."

"I know," I said, "The report always is a lot longer when you find problems."

"Let's have it then."

"First, you were right. When you seal up the back, take out the baffles and close the front door, the top shelf gets too hot right away," I said.

Slim nodded, "OK, but does the fan help?"

"I'll get to that," I said. "Conclusion Number 2: If you open the front door, enough cold air leaks in through the big gaps between the shelves that you don't need a fan tray. The top shelf is cool enough in natural convection."

"Fascinating, captain, but we need that front door!" Slim said.

"OK," I continued. "When I close the front door and turn on the fans, the temperatures do go down. Almost enough. But they don't go down enough to meet the ETSI worst-case altitude spec."

"Altitude?" he said, his mouth dropping open. "ETSI doesn't say anything about altitude!"

I pulled out the dog-eared copy of ETS 300 019-1-3 that I had been reading during those long lab sessions. "Not altitude per se," I said. "It's buried here in Table 1, "Minimum Air Pressure." The spec is 70 kiloPascals, whatever that means. Turns out to be the average air pressure at 10,000 feet."

"No!" Slim cried, pulling at his beard, "As an avid mountaineer, I know that air cooling is less effective at higher altitudes, where the air is thinner. Tell me how much worse it can be!"

"At 10,000 feet, the effectiveness of fan cooling is reduced by about 35%. That means the temperature rise is 35% higher than what I measured in the lab. And when you factor that into my results, the ETSI BK1200 is too hot," I explained.

Slim banged his forehead on his desk. "Tell me how we can fix this, and fast! Or everyone in the facility will become depressed and go on a drinking binge."

"How about a compromise? Remember that with the door open, the temperatures are OK, even without a fan tray? It just so happens that altitude has a much smaller effect on natural convection than on fan cooling. At 10,000 feet, the temperature rise would only increase by about 16% instead of 35%, and that difference, in this unusual case, is just enough to meet the component operating temperature limits."

Slim was skeptical, so I pulled out the good design (GD) guidelines. It says that fan cooling is directly proportional to air density, while natural convection depends on the square root of the air density. Very tricky, that Mother Nature.

Eventually, Slim began to smile. "So the customer saves the price of a door and a fan tray, *and* we get rid of the alarm circuits and filter maintenance, *plus* the system meets the temperature spec? I think I can convince marketing to buy that solution."

To celebrate the successful thermal test, everyone in the facility went out drinking—only, since we were celebrating, it wasn't called a binge; it was a party.

NEBS: THE BIBLE OF THE CENTRAL OFFICE

I get the impression that my colleagues in the telecom industry think of NEBS (Telcordia's *Network Equipment-Building System Requirements,* GR-63-CORE) the same way they think of the Bible. Everybody has a vague idea of what's in it, but it's so big and full of arcane rules that no one actually sits down to read it.

That's why I keep getting questions like these:

"This board is getting so hot I think we might have to use a fan on it. But doesn't NEBS forbid the use of fans in telephone central offices?"

"The hole pattern in the shelf exhaust vent needs to change because we flunked EMI. What hole patterns does NEBS let us use?"

"Our product is justified by faith in its design alone; why does NEBS talk so much about the product demonstrating its good works?"

As Bibles go, though, NEBS doesn't come close to Moses in repetitious laying down of the law. So although NEBS has very little of the imagery and lyrical quality of the Song of Solomon, there's no reason to be scared of it. The whole thing is only 165 pages, and only Section 5.2 mentions fire and brimstone. The thermal requirements take up only about nine pages. Go ahead and read it. Your only excuse may be that Telcordia charges an arm and a leg for each copy, and you don't want to violate their copyright to get a bootleg version. Nobody said working in the telecom industry would be cheap.

But you're the type that waits for the video to come out. So here's my boiled-down version, paraphrased just enough to avoid stepping on the copyright. Just be warned: Don't design your product based on my paraphrase. Telcordia could update the standard at any time, especially without warning the people who have not bothered to buy a copy. But this summary will give you a good idea of what NEBS covers and what it doesn't (see Figure 7-4).

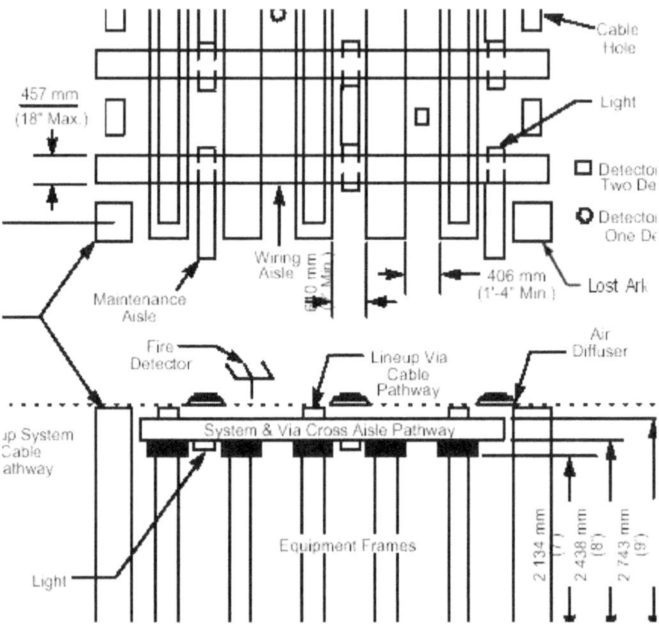

Figure 7-4 *FUN with NEBS.* GR-63-CORE is not as tricky as it looks. Can you find the biblical treasure hidden in this maze of equipment racks and cable troughs?

The 6 Things NEBS Says About Thermal Design

1. Operating Temperature Range: –5 to +50°C

There are many more details about long-term and short-term ranges, and special tests for single shelves, but this is the one you need to know. It's hardly ever 50°C in a central office, but it might be that hot for a few days in a row. That is long enough that your hardware had better be able to carry traffic at 50°C forever. NEBS also gives an operating range for relative humidity, but humidity has almost no effect on component temperature, so I'm going to ignore it. One thing to note is that NEBS applies only to indoor telephone central offices in North America. If you are designing a box for a telephone pole, the basement of a Belgian apartment building or a cable TV station, get the corresponding requirements document.

2. Room Temperature Changes 30°C per hour

Most of the time the room temperature is "normal", around 20°C, so people who work there are comfortable. It only becomes 50°C when the air-conditioning fails. But it doesn't jump from 20° up to 50°C in 10 seconds. It's supposed to take about an hour. This can be important to know. I have seen equipment that worked fine at 0°C, at 50°C, even at 70°C, as long as the temperature was steady. But if the air temperature changed quickly (about 10°C per minute), then it would start hiccupping and taking errors (See Chapter 6.1, "The Milk-box Problem)". Transient temperature mismatches between critical components stopped them from working together, and they wouldn't until the temperature stabilized again. Customers expect to keep passing traffic even when the room temperature fluctuates. So make sure your circuits are not sensitive to changes in air temperature of 30°C per hour. Stop whining! That isn't a very fast change. Imagine designing a car radio that has to start working when it's 40° below zero, and has to keep working as the inside air temperature goes up about 60°C in 15 minutes!

3. Operating Altitude is –197 ft to 5,905 ft Above Sea Level. Operation Required, With "Special Provisions" From 5,905 ft to 13,123 ft

Why do I care about altitude, when I don't care about humidity? Air is our cooling fluid, and at high altitude there is less of it floating around. For a fan-cooled system at 5,905 ft (which is about the altitude in a town called Denver), component temperature rise is about 20% higher than it is in New York City; 20% is hard to ignore. It's even harder to ignore what happens at 13,123 ft, where the scarcity of air molecules causes the component temperature rise above ambient to increase about 50%. Admittedly, there are very few major urban centers in the United States at such an altitude, so NEBS has thrown in the clause about "special provisions." They don't want to force everything to work at 13,123 ft, because very few central offices need it, and the extra cooling equipment would raise costs for every installation. So if you ever get a customer at 12,000 ft, you must document what "special provisions" are necessary to make your product work. Examples of special provisions might be: "Maximum ambient is limited to 40°C" or "Mount rack in chamber pressurized to 1 atmosphere."

4. Average Heat Release Limit: 80 watts/ft^2 Overall. Peak Value: 120 watts/ft^2

That's the limit on the amount of heat a fan-cooled system can dump into the air per square foot of floor space. Yeah, I know, everything that's made these days already exceeds these limits. But it seems that the limits are not going to be increased in the near future, because they are real limits for the majority of existing central office buildings. The customers with lots of existing old buildings are the ones who will be around for a long time, now that all the upstart competitors have gone under and sent their CEOs back to working at Starbucks. Some customers are fundamentalist about NEBS, insisting that these limits are requirements even though NEBS lists them as objectives. I used to think these limits were old-fashioned. Couldn't telecom offices copy the raised floor cooling systems of computer data centers and increase their heat release capacity? A recent study of data centers showed that they are running into big thermal problems with heat release of only 50 watts per square foot! So an "old-fashioned" central office is doing pretty well if it can handle 100 watts/ft^2.

5. Aisle-facing Surface Limit: 12°C Above Ambient

This is pretty simple. Don't burn the customers when they try to open the front door of your electronics cabinet. This has not been a big problem in the past, but it will be as power per rack keeps going up. NEBS lists this only as an objective, but certain customers are "touchy" about it.

6. Acoustic Noise Limit: 60 dBA

When the transistor replaced the mechanical relay in telephone switches, the central office became a very quiet place. Digital electronics just don't make noise. But audible noise is on the comeback trail, and it's all my fault. It's the cooling fans. Listen to one fan gently puffing out its stream of air, and you might wonder what all the fuss is about. But put 24 of them in a rack, and hundreds in a single aisle, and the sound starts to add up. It doesn't take much to exceed 60 dBA. The blower in your home furnace is probably louder than that. So although this is not strictly a thermal requirement, it limits how many watts you might cram into a rack. When I say your board is too hot, maybe, "OK, let's put a bigger fan on it," isn't the answer. That bigger fan might be too loud.

That's it. NEBS in a nutshell. Of course there are lots of other useful things in there that I skipped, like the exact test procedures for temperature and humidity tests, and all the non-operational storage and shipping requirements. Read those after you have committed the six basic rules to memory. Here they are again, in a handy, wallet-size you can photocopy, cut out, and paste on the back of your ID badge.

The 6 Things NEBS Says About Thermal

1. Operating temperature range: –5 to +50°C.
2. Room temperature changes 30°C per hour.
3. Operating altitude is –197 ft to 5,905 ft above sea level. Operation required, with "special provisions" from 5,905 ft to 13,123 ft.
4. Average heat release limit: 80 watts/ft^2. Peak value: 120 watts/ft^2.
5. Aisle-facing surface limit: 12°C above ambient.
6. Acoustic noise limit: 60 dBA.

THE NEW NEBS: MORE A HORROR TALE THAN ANOTHER BIBLE

"Twelve hundred watts in a rack?" I said, perusing the sketch Herbie had slid across the desk at me. "We might even do that without fans."

"Twelve hundred?" Herbie said. He wrote another fat zero at the end of the power estimate.

"Twelve *thousand* watts?" I said. "Are you loco?"

"What's the problem?" he said. "I heard Telcordia put out a new NEBS. It got rid of the 80 watts/ft^2 limit, and allows water cooling. So let's just hook up the cold water pipe to this baby and let 'er rip." (See Figure 7-5.)

I said, "Water cooling! You must have read that on *www.inyourdreams.com*. Why do you need to cram 12,000 watts into a single rack?"

Herbie said, "It's our next generation human brain unit using BWDM—brain wave division multiplexing. By putting a different signal on each brain wavelength (alpha, theta, etc.), each psychic in the system can carry 48 times the traffic of the original HBU."

"Amazing," I said, "but I don't even need to look at the details to say that thermally it won't work, *and* it's going to cause a tremendous power and temperature headache in the central office. And nobody I know is installing water cooling systems yet.

"Plus, there is no 'new NEBS.' NEBS *Physical Protection,* GR-63-CORE by Telcordia) is alive and well and hasn't been replaced or superceded."

Herbie sputtered, "But what about *Thermal Management In Telecommunications Central Offices: Thermal GR-3028,* that Telcordia issued in December 2001?"

"Oh, *that* new NEBS!" I said. "GR-3028 doesn't replace NEBS. For one thing, it only deals with thermal management, and NEBS covers a whole lot more. Even when it comes to thermal, instead of replacing it, it reinforces NEBS and adds a few new things." Then I went into this mandatory impromptu lecture, summarizing it for him.

GR-3028: A Report on "Global Central Office Warming" Disguised as a Requirements Document

Here's what's in GR-3028:

1. A Description of the State of Thermal Management in U.S. Telecom Central Offices Today

During the 1990s, and continuing into the future without any foreseeable slowdown, telecom equipment density, and its power dissipation, have been steadily increasing, with a general disregard for the 80 watts/ft^2 objective in NEBS. Telecom service providers also have been installing data equipment not specifically designed for the telecom central office environment (see Chapter 7.1 for data product airflow problems). Some offices have hit the limits that normal building air-conditioning systems can handle, and non-traditional equipment is making things even worse. Conclusion: Many central offices are on the verge of overheating, and something needs to be done to get thermal management under control.

2. A Standard Language for Describing Room Cooling Classes and Equipment Cooling Classes

Room cooling systems are classified into Overhead Duct, Raised Floor and other categories that you don't care about. Equipment (electronics boxes, that is) is also divided into categories, mostly based on the location of intake and exhaust vents. For example, the HBU Auxiliary Switch Shelf, which draws in air at the bottom front and blows it out into the rear aisle from the top of the shelf, is classified **EC-Class (S) F1-R3.** GR-3028 does not require equipment to be of any particular class, but it does require that you tell your customers what class it is, using this standard language. For a full explanation of those definitions, get a copy of the real GR-3028.

3. Heat Release Targets

NEBS has a single heat release limit: 80 watts/ft^2. GR-3028 tells us that for each Room Cooling Class, certain Equipment Cooling Classes work better than others. If you use the optimum Equipment Cooling Class for a Room Cooling Class, you can safely go beyond 80 watts/ft^2, in the best case all the way up to 150 watts/ft^2. GR-3028 gives a table of Heat Release Targets for various combinations of Room Cooling and Equipment Cooling Classes. Without spelling it out in an official requirement, it does give a strong preference to equipment that draws in air in the front at the bottom, and exhausts it into the rear aisle or at the top of the rack. Because only that Equipment Cooling Class works best with VOH Room Cooling, the type used in more than 95% of central offices. NEBS never spelled that out, but GR-3028 makes it very plain: the front aisle is for cold air, the back aisle is for hot air. If your equipment isn't designed for that, it won't work very well and will probably screw things up for neighboring equipment, too.

4. Accurate Power Reporting Requirement

In the good old days, you could afford to be a little loose in reporting the heat release from a rack. You could just multiply the circuit breaker rating by 48V and say, "It can't be any more than that or the breaker will blow," even though the actual current drawn during the operation of the equipment might be a lot less. Now that the customer is trying to stretch the cooling system to the max, more accurate numbers are needed. GR-3028 requires equipment vendors to supply realistic power values for every configuration, explain how they were calculated and back them up with test data.

Figure 7-5 Even after the release of GR-3028, water cooling in the central office is still a pipe dream.

5. A Reality Check on Air-conditioning Failures

NEBS tells us that when the room cooling goes kaput, the air temperature goes no higher than 50°C, and that it goes up no faster than a mild 0.5°C per minute. The rolling blackouts of 2001 gave Telcordia a chance to witness what really happens in central offices in southern California. In this graph (Figure 7-6), the air temperature increases to 50°C in about 10 minutes and doesn't even think about stopping until after it hits 70°C, and that is with a wimpy power density of only 75 watts/ft^2.

Don't panic. GR-3028 is not going to require equipment to pass a test like this. The NEBS environmental test is left intact, with a maximum of only 50°C. But a fast-ramp test has been added to the end of the old NEBS test. After all the normal cold, hot, high and low humidity cycles have been completed, the product is hit with a temperature ramp from 23°C to 50°C at 1.6°C per minute, to see if the rapid change in air temperature will cause any problems. That is probably not going to make the typical product hiccup. The main headache is that existing large environmental chambers may not be

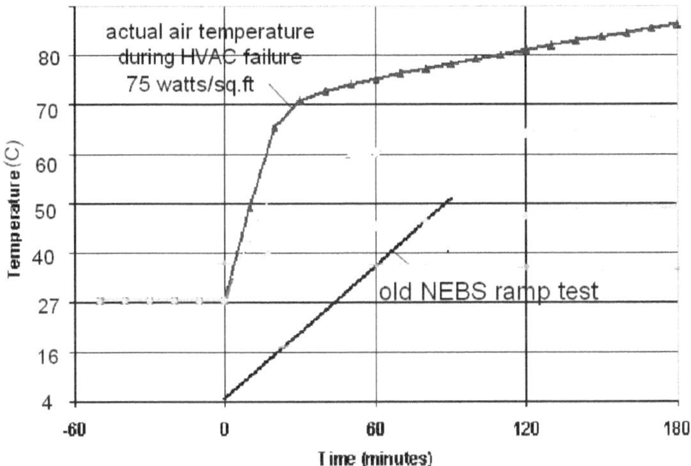

Figure 7-6

able to ramp temperature up that fast. Maybe they could if we put Herbie's BWDM rack in them.

"That's it?" Herbie asked. "All we have to do is ramp up the chamber a little faster, tell which end the air blows out of and report the power dissipation? Where's the water cooling?"

I flipped through the pages of GR-3028. "It does mention water cooling. It speculates that if power keeps going up, central offices might be forced to adopt water cooling, but it doesn't have any requirements about it. Oh, it does say that if your equipment uses liquid cooling, it should be listed as EC Class LQ."

Herbie was crestfallen. "This new NEBS doesn't do me any good at all then. What about the BWDM rack?" Herbie asked.

"Until a water cooling standard does come out, I suggest you scale back from the impossible 12,000 watts to the merely outrageous 5,000 or 6,000 watts per rack. And I'll start brushing up on my plumbing skills," I answered.

NORMAL ROOM TEMPERATURE: THE LATEST WORST-CASE THERMAL CONDITION

When Herbie set the cardboard box on my desk, it made the cubicle walls creak. From a sea of foam peanuts, he pulled out a fan (see Figure 7-7). "This is the answer to all our problems in Point 2," he said with his usual exuberance.

"Point 2" was our shorthand for HBU feature package 8.3.7.2. So many software patches had accumulated for this release that we now had to retrofit a more powerful processor onto all the boards to handle them. The power dissipation of the processor would increase from 4.5 watts to 47 watts. I was struggling to conceive of a heat sink that could do the job and fit in the existing space (none, since the previous processor did not need a heat sink).

Herbie's fan looked innocent at first glance, the same size and shape as the 5-inch-square fans we already had in our fan tray. But this one was much heavier, with a beefy metal frame and heavy gage power wires. "Are there *gussets* on the fan blades?" I asked.

Figure 7-7 Fan vendors are learning to make more powerful fans to deliver more air flow in the same space, but the flow you can use is really limited by the NEBS audible noise limit.

"Oh, yeah!" he said, giving the rotor a spin with one finger. Even spinning unpowered, it sounded like a food processor slicing into a raw potato. "Without gussets, at full speed the titanium blades would start to warp from the combination of backpressure and centrifugal force."

"No kidding?" I said. "So you think this fan will give us enough air flow to fix the Point 2 processor?"

"Maybe. You said if we could double the air flow and then add the biggest possible heat sink, you might be able to get the processor under its operating limit. The spec on this new fan says it has double the flow of the old one, and it drops right into the existing fan tray," Herbie said.

"When I said double the flow, I assumed it would be impossible and you'd give up," I admitted. "But it's worth a try. What's the audible noise on this beast?"

Herbie squinted at the spec sheet and read, "*'Does not cause much hearing damage in laboratory mice.'* Must be OK. Why do you ask?"

"You do realize, based on our adventure in Chapter 2.7, the audible noise of a fan goes up really fast as you increase the fan speed, right?

To double the flow rate, this fan would have to double the RPM, assuming the same blade design. When you double the RPM, the audible noise goes up about 15 dB. That may not seem like a big deal, but a 10 dB increase sounds about twice as loud," I explained. "And at full speed, our old fan tray was already over the NEBS noise limit."

"That's OK," Herbie said, "We'll use the same trick as the current fan tray. When the room air temperature is below 40°C, we slow down the fan until it meets the noise spec. We really need that huge amount of air flow only when the ambient is 50°C, or during a fan failure, right?"

I nodded and took the fan spec sheet from Herbie. "Let me plug this new fan curve into my Therminator thermal simulation model of the shelf and see what we get at 50°C," I said.

When the simulation work was done, Herbie asked me to present the results directly at the HBU Central Command Meeting, so everyone could see how he had saved the project. "Maybe you better hear what I ended up with first," I suggested.

Herbie, although not the sharpest crayon in the box, could smell trouble. He snuck over to my office during lunch.

"What's the big deal?" he asked. "Didn't that super-duper fan help?"

"In fact, it did!" I said, "Your discovery of this monstrously powerful fan, coupled with my design of a highly optimized bonded-fin heat sink with phase change interface, has produced a processor temperature that is 5°C under its operating limit, even at 50°C ambient, with one fan failed in the fan tray."

Herbie grinned and held up his hand for a high-five. I raised my hand in a similar manner, but held it more in the way a traffic cop signals cars to stop. "Not so fast on the celebration," I said. "I didn't give you the results at the worst-case condition."

"Worst case?" Herbie said. "What could be worser than 50°C with one dead fan?"

"I didn't realize it myself until last night," I said, "but for this particular fan tray, it turns out that the worst case happens when all fans are working, with an ambient of 40°C."

"40°! How can temperatures be higher at 40°C ambient than at 50°C? Especially with all fans working?" he said.

"Because at 40°C, we still try to meet the NEBS audible noise requirement, since 40°C is considered a long-term, normal operating condition for the system," I explained. "So at 40°C the fans all run at a much slower RPM, the same RPM that our old fans had to run at to achieve 60 dBA. When fans with the same blade design spin at the same RPM, you get the same audible noise and the same air flow. So even though we have put in this new gale-force fan, when we run it at low speed to meet the audible noise spec, we get only the same air flow as we did with the wimpy original fan."

"But the ambient is 10°C lower," Herbie tried. "Doesn't that make up for the slower speed?"

"It makes up for exactly 10°C of the higher processor temperature," I said. "In this case, it just isn't enough. When the air flow is limited by audible noise, the processor is going to have a junction temperature of about 130°C, even with the amazingly efficient heat sink I designed for it."

Herbie slumped down into a chair. "How about if we change the trigger point in the fan tray so it goes to full speed at 30°C ambient instead of 40°C?"

"OK," I answered, "then at 30°C ambient the processor will be 120°C, which is still 15°C over its operating limit. If you want to solve the problem that way, you'll have to start running at full fan speed at 15°C ambient, which is below normal room temperature. (Normal room temperature is about 18 to 20°C.) That means the fans will violate the audible noise spec practically 100% of the time, which the customers won't appreciate."

Herbie picked up the monster fan and flicked at the blades as he thought. I couldn't tell if the grinding sound was coming from the fan or from his brain. Eventually, he pinched his finger between a blade and the housing.

"So the main (*ouch!*)—what do you call it?—moral of this tale is: We could keep increasing the heat density if we could keep increasing the air flow, but the air flow is limited. Not because fans can't do it, but because we eventually hit the audible noise limit," he said, sucking on his bleeding finger.

"Unless the audible noise limit in NEBS changes, we're stuck. Of course, you *can* get more flow with relatively quiet fans—if you increase the diameter of the fans, that is—and make them a *lot* bigger. But I thought the whole point was to fit more electronics in the rack, not more cooling equipment," I said.

Herbie nodded seriously, "OK then, I guess we have no choice. I'll have to resort to my last resort to make this puppy work thermally."

"Sounds like you're desperate," I said, "which is not when you get your best ideas. Shoot."

"The software team confided to me that they only really need the maximum computing power of this processor when it is running through their bumpiest, patchiest chunks of code, which doesn't happen very often. They say it only runs about 10 or 15 seconds at max power every couple of hours at the most. The rest of the time the processor is mostly idle, at about 4 watts. Does that help at all?" he offered.

"So you heard about that melting-core heat sink (see Chapter 6.6 for that and other crazy ideas that just might work). Maybe that, or just a big, heavy sink with lots of thermal capacity, will help. The processor power is mostly 4 watts, but 47 watts for only 10 seconds once in a while. That averages out to about 4 watts, which is what the old processor was. You should have told me about this in the first place. Then we wouldn't have needed this new fan and you wouldn't be needing stitches in your finger. Come on, I'll drive you to the emergency room. Again."

THE WEAKEST LINK IN AIR COOLING

So far this 21st century has been a big disappointment (Figure 7-8). I grew up on the science fiction of the 1950s and '60s. Plus, I knew that the stories of Jules Verne and H. G. Wells had turned out to be amazingly accurate predictions of real technological advances. So I was expecting that if I lived long enough to see the 21st century (no way—I'll be in my '40s!) I'd be living in a technological paradise, or a post-apocalyptic nightmare or at least in a dull, Big Brother–controlled pseudo-utopia. By 2001 there would be routine space travel, telepathy machines, personal jet packs and—most important of all—robots to do all the work. Well, Y2K+1 has come and gone, and technology has not kept up its end of the bargain. I can easily live without the silver suits with the big shoulders, but where are my flying car, my jet pack and my house-cleaning robot? Government is making headway with the Big Brotherly oppression, but it hasn't even broken ground for the dull pseudo-utopia. The technology we did get—cell phones and video recorders—are merely small improvements on things I could already do when I was a kid: talk on the phone and watch TV. It's time for the robots!

Figure 7-8 "21st Century—No people allowed."

It's the 21st century, and we're still using air as the cooling medium for most electronics. I'm not complaining about that. Air is still being used because it does the best job for the money, not because we haven't invented anything more high-tech.

But these days it seems that for large electronic systems, like telecom and data processing equipment, the heat density is so high that we are reaching the limits of air cooling. Not that we can't blow enough air through an electronics chassis to adequately cool the circuits inside, although that day is not far off either. What I'm referring to is that when you load up a room with network servers, there is so much heat that the room air-conditioning system can't handle it. If you can't remove all the heat that is generated in the room, then you can't control the room air temperature, and the electronics will eventually overheat. I didn't make up this trend to suit my science fiction theme. It has been well-documented by The Uptime Institute (although it sounds like something I might make up, this is a real organization with real data).

Various sources report different limits for room air cooling. They range from about 540 watts per square meter (W/m^2) of floor space up to maybe 2,700 W/m^2 (about 50 to 250 W//ft^2). At the low end, the limits can be blamed on old buildings with undersized air-conditioning systems that can't be upgraded or the fact that the cooling system is poorly designed to handle dense heat loads. But there is rarely a detailed explanation of the upper end limits. I have already worked on telecom systems with heat density of 3,000 W/m^2 and up. How can I sell products that will overheat my customers' sites?

What is the reason behind the limits of air cooling? Is it the wimpy air-conditioning fluids we are stuck using these days to save the ozone layer? Is there some physical property of air that poops out above a certain temperature? Is it government regulation? In other words, who can I blame (besides myself) for data centers and telecom offices on the verge of overheating?

I try to solve every problem by doing an energy balance. In most personal situations it doesn't help much, but for getting the heat out of a room, it is pretty good. Let's start by saying we have a room 6 m wide, 6 m long, with a 6 m ceiling. One duct brings in cold air from the air-conditioning system, and a second duct takes away the hot air. Someplace outside the room they meet at the air-conditioning plant. The amount of air coming in equals the amount of air going out, ignoring any leaks (see Figure 7-9).

It is pretty easy to write an equation that tells us the amount of heat that can be removed from the room by the air cooling system.

$$\mathbf{Q = A\ V\ \rho C_p\ (T_{hot} - T_{cold})} \qquad \text{Eq. (7-1)}$$

where $\mathbf{Q}$ is the heat removed from the room, $\mathbf{A}$ is the cross-sectional area of the duct, $\mathbf{V}$ is the air velocity in the duct, ρ is the density of air, $\mathbf{C_p}$ is the specific heat of air, $\mathbf{T_{hot}}$ is the air temperature leaving the room and $\mathbf{T_{cold}}$ is the air temperature entering the room.

I know the density and specific heat of air. They don't change much over the range of temperature we normally deal with, and even when they do, I can look up their values. But what do we plug in for the other parameters of the equation?

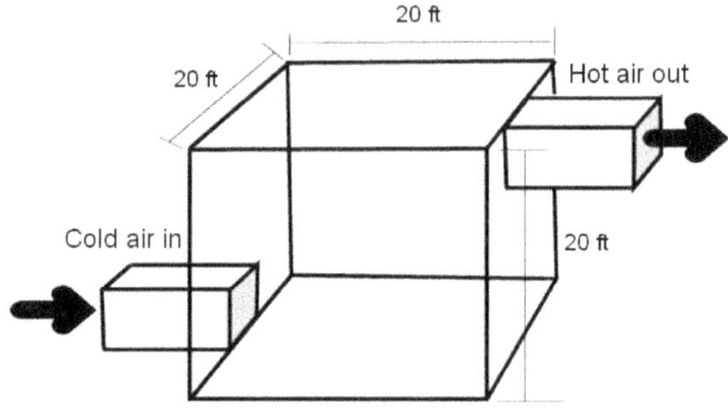

Figure 7-9

Let's start off with an ordinary size for an air duct—0.5 m². That is actually pretty big, compared to the ducts in your house. So the duct area *(A)* is 0.25 m² as a starting point.

Then we have to pick an air speed for the inlet and outlet vents for the ducts, and an inlet and outlet temperature. Let's assume we have an infinitely powerful air-conditioner and blower, perhaps the unit that makes the Superdome in New Orleans comfortable, so we can choose any velocity and temperature we want. What else limits our choices, besides our lack of imagination?

One not-so-obvious limit on the cooling system is that data centers and telecom offices employ human beings. The people and the electronics are made comfortable by the same room cooling system. The air temperatures and velocities have to be kept in the narrow range that allows for human comfort.

I am not a human comfort expert, but I have chosen some values that seem right to me. An HVAC engineer might quibble with my numbers, but I think they are in the right ballpark.

Maximum duct velocity: 2.0 meter/sec (4.5 miles/hour)
Minimum inlet air temperature: 15°C (59°F)
Maximum exit air temperature: 28°C (82°F)

This air speed and temperature range is not going to kill anybody, or even give them the sniffles, but I think you'd get annoyed if you had to sit at a workstation all day with a 2 m/sec wind at 15°C blowing on the back of your neck, or had to work on cabling at the top of a cabinet in 28°C air.

Plugging these values into Eq. (7-1) (density = 1.16 kg/m^3 and specific heat = 1,000 J/kg°C), I get a heat removal capacity of about 7,500 W. The floor is 6 × 6 m, which is 36 m^2. A room with this cooling system can have a heat load of only 210 W/m^2. Is that all? How can we get our room up to something reasonable, like 1,000 W/m^2, or preferably 3,000 W/m^2, so I can sell my product?

One thing we can do is add more ducts, or make the ducts bigger, to get more total air flow, without making the duct velocity higher. Table 7-1 shows how many ducts we would need to increase the heat removal capacity to a desired level.

The last line in Table 7-1 is the physical limit for our room. If we made the left wall into one giant inlet duct and the right wall into one giant exit duct, and connected that 6 × 6 m duct to our infinite air-conditioning system, it could sustain a heat load of about 30,000 W/m^2. To do that, the whole building would become a wind tunnel, which, although not impossible, would compromise most of the other functions of the building, such as interconnecting data lines.

So let's back off from that extreme, and say that only half the left and right walls are occupied by ducts, so there is still some wall space left to hang occupational safety posters. That would support a heat load of 15,000 W/m^2. That is the type of room you need to support the 10,000 W/m^2 equipment being introduced these days.

Table 7-1. Air Cooling Limited by Human Comfort

No. of 0.5 × 0.5 m inlet ducts	Heat removal capacity
1	210 W/m^2
2	420
3	630
4	840
Entire wall open (144 ducts)	30,000

This gloomy scenario hinges on one important assumption—that we need to worry about human comfort in the room. What if we could kick the humans out, and set the air velocity and temperature range only for the safe and reliable operation of the equipment in the room? What happens if we pick the following values?

Maximum duct velocity: 10 meter/sec (about 22 miles/hour)
Minimum inlet air temperature: 5°C (41°F)
Maximum exit air temperature: 50°C (120°F)

I chose 5°C so we don't have to deal with frost formation. Otherwise we might be safe at 0°C or even a little lower, assuming that humidity is controlled to prevent condensation. Telecom equipment would thrive in such an environment, but your maintenance crew definitely wouldn't stay in that room very long. Table 7-2 presents the heat removal capacity for such a room:

If you ignore human comfort (or even survival), the heat removal capacity instantly increases by a factor of about 17, and maybe more. If we eliminate people from the equipment rooms, then the only thing we'd still need to do is to find out where the Superdome bought its infinitely large air-conditioner.

It turns out that the weak link in the room cooling system is the presence of humans. This is where the robots of the "21st century That Should Have Been" come in. If we had been working on robots to do the maintenance in our telecom offices, data centers and computer rooms, we could be extending the limits of air cooling for elec-

Table 7-2. Air Cooling Limited By Equipment Operation

No. of 0.5 × 0.5 m inlet ducts	Heat removal capacity
1	3,600 W/m^2
2	7,200
3	11,000
4	14,000
Entire wall open (144 ducts)	520,000

tronics right now. Of course, we should make sure to include in our robots a suitable remote control "off" button, for that inevitable moment when they realize humans are the weak link in many other realms as well.

This chapter was originally published as an article in *Electronics Cooling* magazine, May 2003, Vol. 9 No. 2. It is reprinted with permission.

INDEX